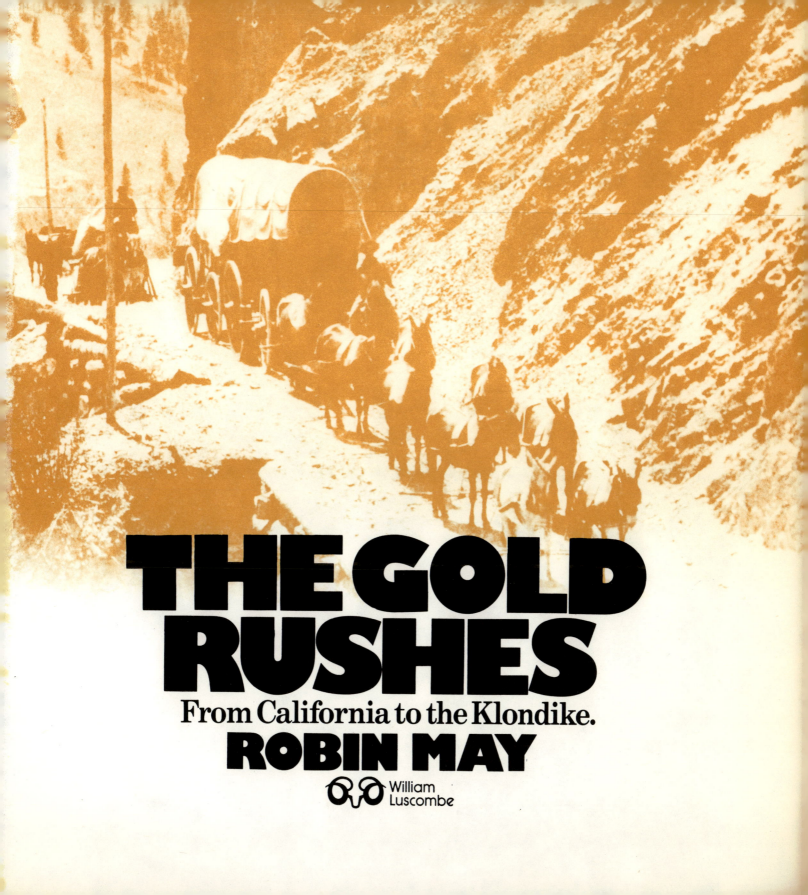

THE GOLD RUSHES

From California to the Klondike.

ROBIN MAY

William Luscombe

THE GOLD RUSHES

Printed in Hong Kong by Mandarin Publishers Ltd.
Phototypeset by Tradespools Ltd, Frome, Somerset, Great Britain
Designed by Grass Routes, Marshfield, Avon

First published in Great Britain in 1977
by William Luscombe Publisher Ltd.
The Mitchell Beazley Group
Artists House, 14–15 Manette Street,
London W1V 5LB

ISBN 0 86002 152 1

To
Joseph G. Rosa
with whom I saw Deadwood
and the Black Hills of Dakota
and to
Sue Herzog,
Everett De Golyer
and Jim Phillips
who showed us round them

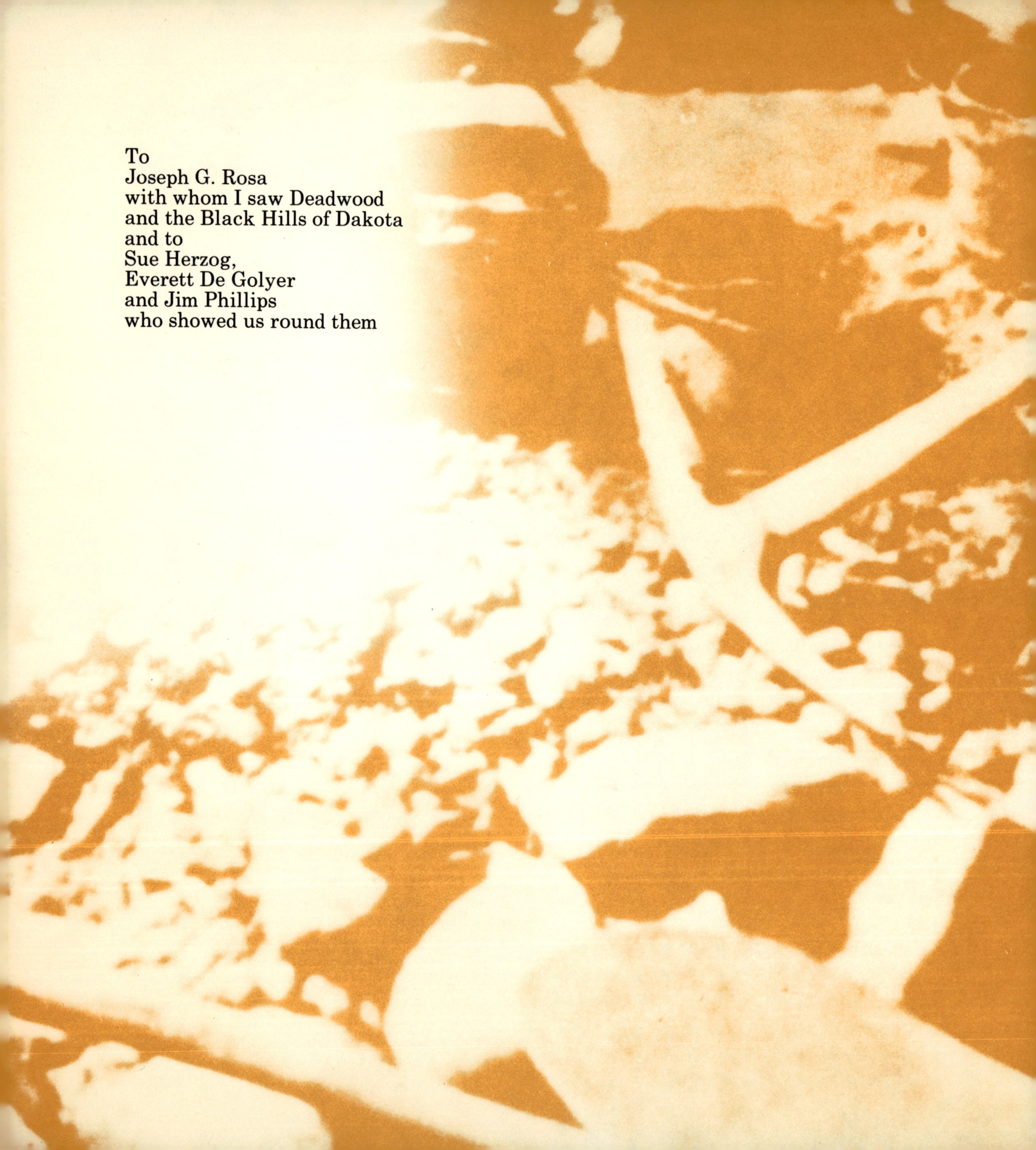

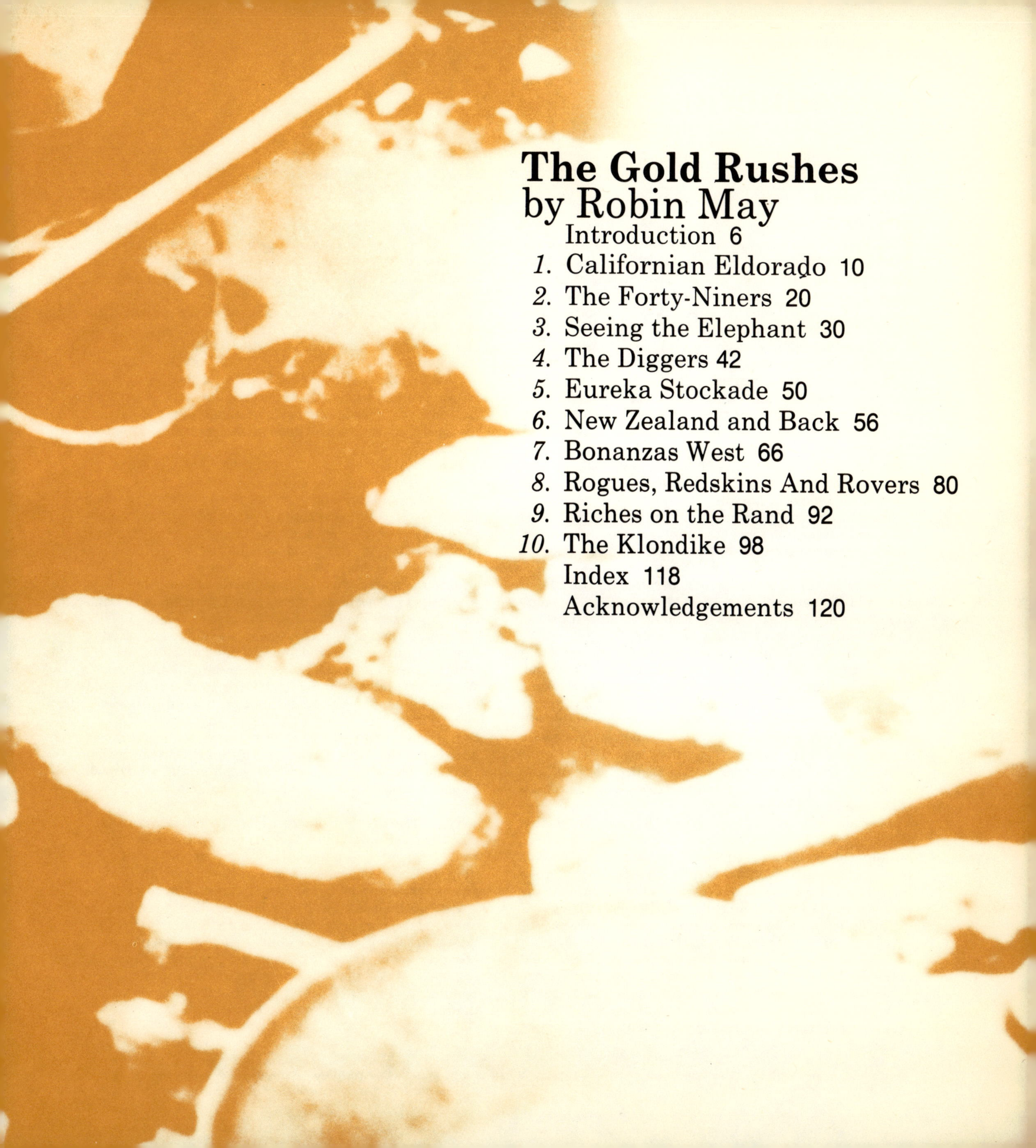

The Gold Rushes
by Robin May

Introduction

'Gold! Gold! Gold on the American River!'

This was the cry, so thrilling to the citizens of the tiny village of San Francisco that May day in 1848, which started half a century of gold rushes. A travel-stained, extrovert Mormon named Sam Brannan kept shouting it as he paraded the streets, waving a bottle full of gold dust to prove his bold claim.

Man's relationship with gold was already a long one. The precious metal had been known over much of the Middle East some 7,000 years earlier and had been used in considerable quantities by the Egyptians. Since then, and since Sam Brannan's victorious shouts, it has been continually sought, shaped, cherished, lusted after and fought over. The New World in particular proved a gold-seekers' paradise, with the Spaniards in the van, taking first Montezuma's treasure, next the fabulous hoards of the Incas, then seeking Eldorado.

This was originally a man, *El hombre dorado*, the Gilded Man, an Indian chief in what is now Colombia, who, when elected, plunged into Lake Guatavita, his body smeared with gum and gleaming with gold dust.

Gradually, *El hombre dorado* became El Dorado–or Eldorado–a mythical city of gold, to which so many thousands were drawn, but which none found; and a word, rich in overtones, entered many languages.

This book begins with the supreme rush to California that set the pattern for those that were to follow. There had been earlier rushes, but none had been on an international scale. Brazil had had one from the late 17th century to early in the 19th, an exceptionally long one in cruelly difficult country, where primitive versions of later techniques for getting gold were learnt. The effect was a population explosion, from some 300,000 in 1700 to around 2½ millions in 1800, including thousands of Portuguese and hundreds of thousands of African slaves.

Siberia, too, produced gold in the 18th and, especially, the early 19th centuries, but not until the 1930s did modern mining methods, long used elsewhere, sweep the U.S.S.R. into second place (after South Africa) among gold-producing countries. Stalin is said to have encouraged the search for gold after reading about California's rushes.

The first American rush occurred in the South, not the West. A nugget had been found in North Carolina by a boy in 1799, but it was in northern Georgia in 1828-29 that major strikes occurred. The ensuing rush is best remembered today because of the wretched fate of the peaceful and advanced Cherokees, whose land was brutally grabbed, despite a Supreme Court ruling to the contrary. President Andrew Jackson, ex-Indian fighter, and strictly Frontier in his attitude to Indians, sided with the gold-hunters.

While the Cherokees were driven westwards along their 'Trail of Tears', their lands were overrun, but though the finds on them and elsewhere were reasonable, the true value of the rush was that a number of miners learnt the skill of alluvial mining, later using this skill in California and elsewhere. Alluvium is the sediment of streams and floods.

Now at last the stage was set for rushes on an international scale, rushes that had young men in New York, Boston, Sydney and London leaving their jobs and heading for the docks–rushes that depopulated whole towns within hours.

This book is about people, those early on the scene, the individual gold-seekers, and the thousands that followed them, heading for, and just occasionally finding, their Eldorado. It is not about the various mining techniques, though the key ones are described, and it is not solely about miners, but everyone from merchants to con men and gamblers, from dauntless pioneer women to often equally dauntless 'soiled doves', from native-born Americans, Australians, New Zealanders, South Africans and Canadians to citizens of most of the countries on earth.

Because the book is about people and not gold production, the amount of gold found does not determine the length of each section. By any standard, South Africa's Witwatersrand was and remains the most amazing gold-producing area in the world, but with capitalists, made rich by diamonds, bringing the most advanced machinery to the Rand, and with the most exciting spot in Johannesburg being the Stock Exchange, the South African rush of 1886 had less of the extraordinary colour, of the anguish, hardship and outsize individuality, than most of the other rushes.

Today, the great gold rushes have undeniable entertainment and escapist value, and in their own day they were rightly regarded as phenomenal and thrilling events. But they were also choice examples of the human spirit, mad and magnificent in equal proportions, a view that the author of this book has firmly held since first reading the one modern classic about a gold rush, Pierre Berton's *Klondike*. They were certainly mad, for as one early observer noted, 'the only occupation in which men seem to engage without the least preparation or forethought is mining.' As for their occasional magnificence, it will be the author's fault if this is not apparent.

R. M.

'More gold' trumpeted the *Alta California* nearly two years after the first finds, while the Pony Express rider is spreading the good news of yet more finds a decade later.

Alta California.

E. C. Kemble, J. E. Durivage, } EDITORS.

TRI-WEEKLY.

TERMS

VOL. I. SAN FRANCISCO, WEDNESDAY MORNING, DECEMBER 19, 1849.

THE
TRI-WEEKLY ALTA CALIFORNIA
IS PUBLISHED
Every Monday, Wednesday and Friday,
BY E. GILBERT & Co.
E. GILBERT, E. C. KEMBLE, R. C. MOORE, J. B. ORMISTON.

TERMS.
For one year, in advance, per mail $16.00
For six months, " " 10.00
Single Copies .... 12½ cts.

TERMS OF ADVERTISING.
One square, 10 lines, (or less) first insertion $4.00
" " each subsequent insertion 2.00
PAYABLE INVARIABLY IN ADVANCE.

BOOK AND JOB PRINTING
Of every description neatly and rapidly executed at the Alta California Office.
Office on Washington st., Portsmouth square

Tri-Weekly Alta California.

WEDNESDAY MORNIN

"More Gold."

From the memorable disco
mountains of the American
more
ing substa
States, from Maine to
pers can well inform them
While *about one-half* of the
ries were gravely declared di
scrupled to swallow *all* perta
very of gold somewhere in
Mountains, or away down east
far-west; but always, it wou
a vicinity plowed over for half
Even to this day, the
heads many scathing
followed by the plau
gold" on the bord of Arka
may be. If thi e not "s
swall nel," pe
the pa with more truth.
Texas papers keep up the
Witchita Mountains. A go
have gone there, and hope
respectable emigration
coming summer, to t
road. An eastern paper, quoting the
tinues:

We take from the reports which the papers contain some paragraphs which will serve to feed the appetite for the shining metal.

A correspondent of the Fort Smith Herald, writing from the Chickasaw District, Indian country, on the 26th, says:

A man came from the Witchita Mountains, a few days since, with a cactus, the *dirt around the roots was washed, and proved to be of such value that it fully compensated him for the trouble of his journey*—it has set all northwestern Texas on the

gings and washings on the land of Mr. Elisha Thurman, in Nelson county, *which was lately advertised in our paper for sale, and we believe is still in market.*

One day, four hands engaged in washing for gold, procured 163 pennyweights, worth about one hundred and sixty dollars; and, in the same manner, within a short period, some thirty or forty thousand dollars has been obtained at very small expense. Strange to say, there is in the same neighborhood, a rich lead mine, which yields eighty per cent. lead, and four dollars of *silver* to the one hundred pounds.

We hope our Richmond cotemp
don our intimation that this stor
little *steep*. We think the privilege
it quite becoming to us, who are
flat contradictions of California
when in reality we are
statemen

There is no
his doom—no portion of the castle is breached, but the vice is turned, the valves fly apart, blinded and bewildered, the live oyster is consigned to tickle your palate with his dying agonies."

SEVERITY OF PRUSSIAN MILITARY DISCIPLINE.—A correspondent from Randers, in Northern Jutlan, dated the 30th of June, has the following:

On Monday last, when the 2d battalion of the 18th regiment of the Prussian Landwehr which is garrisoned at Randers, was ordered to attack a

Interesting Communication.

We have to acknowledge the receipt of the following communication from Mr. B. Buggins, the gentleman who received one vote at the recent election, and to whom we suggested the propriety of establishing a neutral restaurant under the title of Buggins' Retreat. We cannot conscientiously longer keep our readers from the perusal of this singularly graphic effusion.

SAND FRANCISCO, Dec. 17, 1849.

MESSURS EDYTURS:—I see a peace in your paper the other day containing a honorable menshun of my name an a allushin to my receiving a e at the recent election. I am no aspirant for politikel honors and my name was placed upon the ticket without my nollidge as was my friend Muggins's wich accounts for the fact of my only receivin one vote. I have so much con ence in my friends as to that if the Buggins banner Buggins would Polytix how- blow the han be a eatin no tom themselves neu- the city would only lease square should make lots be the least thing e a man to put bread into oples. The town ain't at t *selling* property, though e money there aint one in n't like to rail agin 'em, hey pays much attenshun onpleasant to them as s no house-keeper but so very bad that she ed to keep house in s it very difficult for o' a very temperate e all got pin or they all go set well on their esses it will be going out children is weter in the way f mischief, and the chief of the on em is Benjamin, jr.

Mrs. Buggins and me meant to pay visit but thinkin that you might not like to see air on us, and that our visit might be fruitless couldnt bear to trouble you knowin how m eddyturs is troubled by visiters from havin be in the literary way myself, havin once had the charge of a literary depot. My friends used to come in and gobble up all the light literatoor until the whole concern literally went to pot and every thing was knocked into pie, a perfect pot pie. My returns was very slim and I forsook the paths wich is strewed with roses and briars and rose up with a determination to stick to my old trade, wich is that of a carter so I shall think no more of poetical cartoones specially since Mr Poe

THE MEN TO MAK AND THEIR MARKS— title of an address b trustees, teachers a College, on the last a Independence. The sages:

And for the marks State. *I see them in* right at you, with his that mantles on his ch it is but virtue's colo You know that boy i The darkness is not c lightens it. He is m wrong, and readier for There is nothing poss count on him; and n sible to him and God.

I see them in the ea athrob in all his hand fixes on the page of Ho and never wavers till has stored its treasur mind. His foot has nevolence or mercy. bounding ball fly hig the stand, and hear t signal of his triumph, was his strong arm t

I see them in the re where others stand. when superiors pass, by. He owns in ev prayer, a present Goo erential boys; these make a State.

"What constitutes a
Not high-raised battle
Thick walls, or moa
Not cities proud, with
Nor bays, and broad
Where, laughing at th
Not starred and spa
Where low-browed ba

Men, who their duti
But know their rights
Prevent the long-ai
And crush the tyrant,
These constitute a S

HUN WOM in Hungary ive dence of the bravery thing in the Amazo tions. The *Europea* ing description of the Patriotism and tr eat characteristics Hungary. Ladies well as those of th gled together in a stand forth as encour Republic. The you foremost in the blood giment of volunteers actully in command a lady of rank, and i the uniform of officer

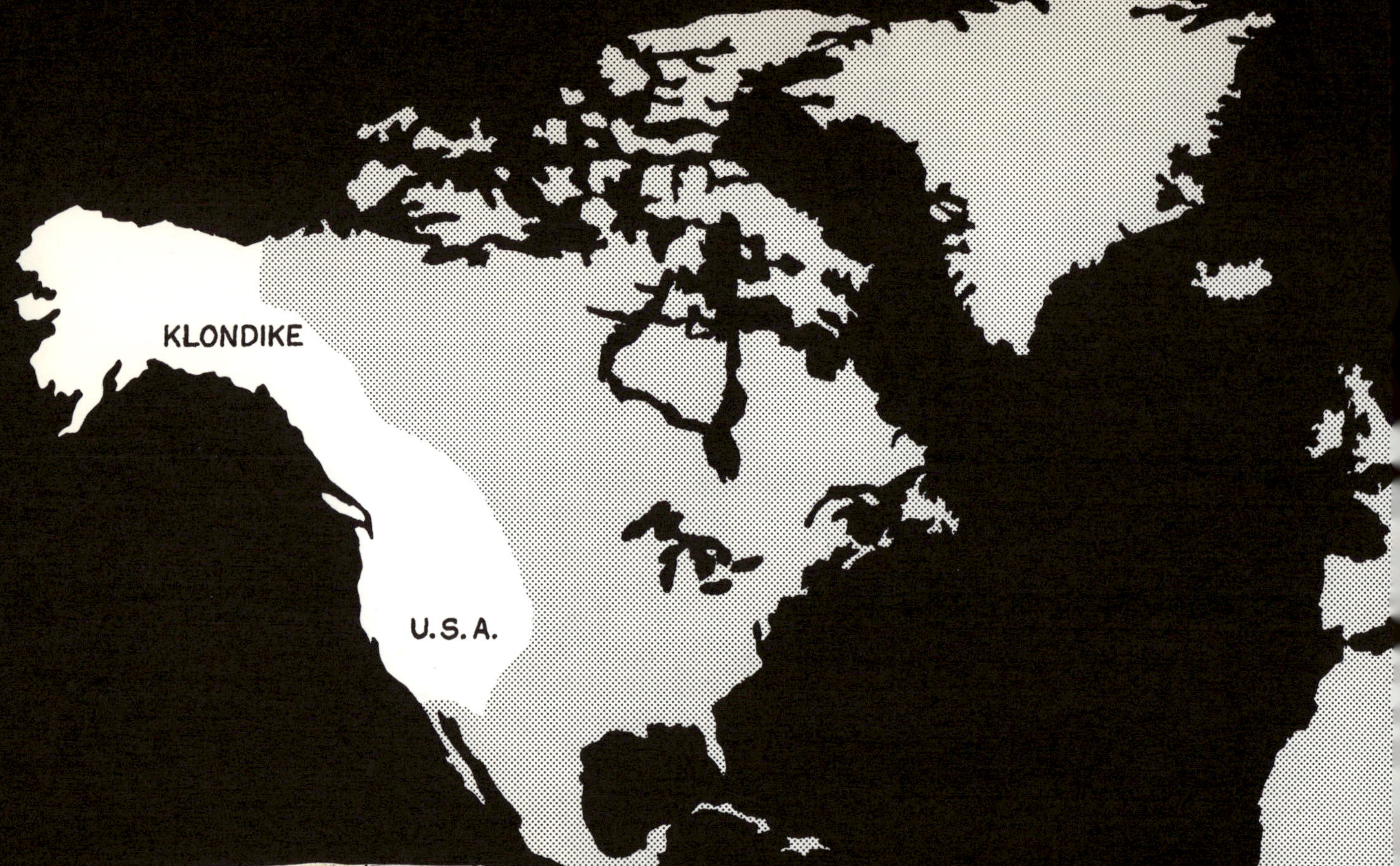

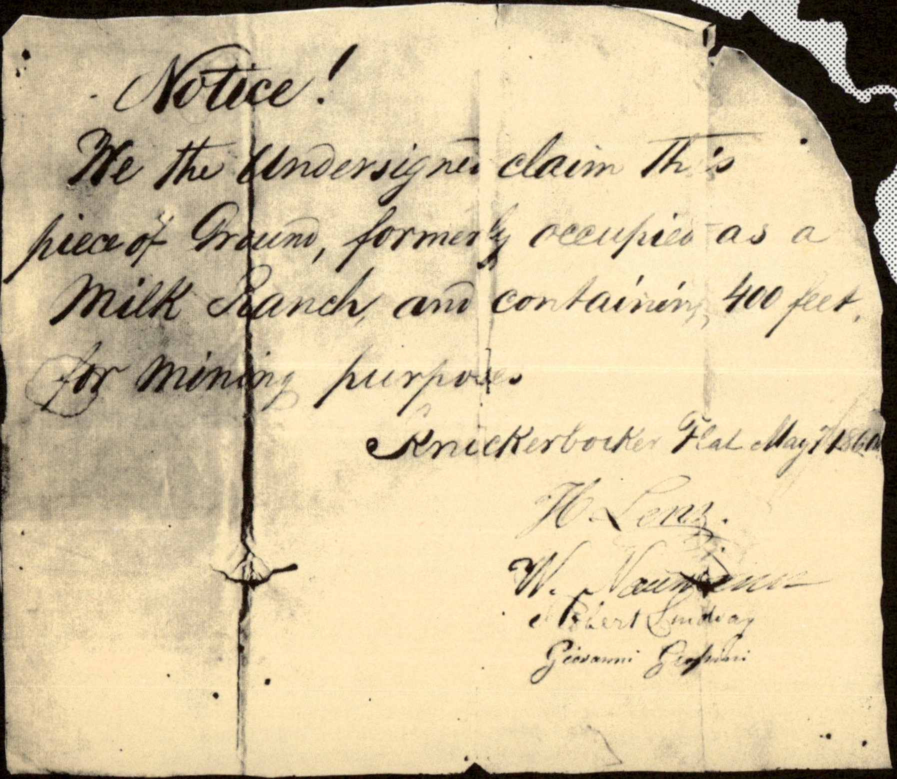
Notice!

We the Undersigned claim this piece of Ground, formerly occupied as a Milk Ranch, and containing 400 feet, for mining purposes.

Knickerbocker Flat May 7 186[illegible]

H. Lenz.
W. [illegible]
Robert Lindsay
Giovanni [illegible]

Over a map of the gold-mining areas of the second half of the 19th century is the quarter-ounce nugget that convinced James Marshall he had struck gold, also a notice showing how simple it was to stake a

California Eldorado

Eureka! Oh how my heart beat! I sat still and looked at it some minutes before I touched it, greedily drinking in the pleasure of gazing upon gold that was in my very grasp. . . . E. Gould Buffum: *The Gold Rush*

John Sutter was virtually ruined by the finding of gold on his land

In his private fort on the Sacramento River, 'Captain' John Augustus Sutter was enjoying the good life. Born in Switzerland, he had fled across the Atlantic in 1834 to escape from his creditors, finally settling in Mexican-owned California in 1839.

This remote, neglected, vast outpost of the old Spanish empire had had a comparatively peaceful history, in striking contrast to the savage wars between Indians and Spaniards in what is now the American Southwest. Only in California did Spain's dream of a colonial empire succeed and deserve to succeed, albeit in a modest way, with a devoted number of priests establishing missions in the 18th century, converting the peaceable local Indians and putting them to work. A few government officials and soldiers completed what was a by no means unpleasant pastoral scene, worlds away from the stark horrors of the Southwest, where Apaches and Comanches took ferocious toll of the grasping heirs of the iron Conquistadores.

The gold-crazed Spaniards never seemed interested in California, and even after Mexico became independent of Spain in 1821, little changed. A few small gold strikes in the early 1840s were mainly notable for the total lack of excitement they caused, at Sutter's Fort or anywhere else.

By now the Oregon country to the north had become American, enough settlers braving the 2,000-mile journey from the Missouri to win a bloodless dispute with Britain by right of physical presence. A few Americans had already headed due West for California, and in 1847, during the war with Mexico, the province was slickly annexed. The timing was immaculate, for gold was to be discovered on January 24, 1848, just over a week before the official peace treaty finally made California American.

The twin events should have been the making of Sutter, who was already a major figure in California, but the very opposite occurred. He had prospered mightily after becoming a Mexican citizen and receiving a 50,000-acre land grant. His riverside fortress home was the headquarters of a fertile, feudal kingdom which he named New Helvetia, even though everyone called it Sutter's Fort. In it he installed his two Hawaiian mistresses, and beyond it a mainly Indian work-force manned his domain. Yet his prosperity was more apparent than real. His workers were unskilled and many were idle, his foremen were mostly inept. He was almost too generous to immigrants and travellers, and his debts began to spiral. However, with such a land-holding the Captain (who had given himself his rank for imaginary services rendered in the Swiss Guard of King Charles X of France) would surely have struggled along very pleasantly. Then his partner found gold.

This partner was James W. Marshall, a born loser, for all that he made the most important single gold strike in history. A carpenter in his early thirties, inclined to fits of gloom long before his total failure to make anything of his find, he teamed up with Sutter on equal terms to build a sawmill.

Marshall chose a spot on the south fork of the American River, some 45 miles from the fort. It was not on Sutter's land, but belonged to the Coloma Indians, who were not consulted. The combination of good timber and enough water to power the mill made it an admirable spot.

Marshall employed Sutter's Indians and some Americans, including members of a Mormon Battalion who had reached California too late for the Mexican War and been disbanded.

When the mill was almost complete it was found that the tailrace below the wheel was too shallow to turn it fast

enough, so digging restarted and each night water was allowed to flood through to clear away as much sand and gravel as possible. And each morning Marshall would walk down to the mill, shut off the water, and see what progress had been made. His own own words vividly recall what happened on the 24th:

> One morning in January–it was a clear, cold morning; I shall never forget that morning–as I was taking my usual walk along the race after shutting off the water, my eye was caught with the glimpse of something shining in the bottom of the ditch. There was about a foot of water running then. I reached my hand down and picked it up; it made my heart thump, for I was certain it was gold. The piece was about half the size and of the shape of a pea. Then I saw another piece in the water. After taking it out I sat down and began to think right hard. I thought it was gold, and yet it did not seem to be the right colour ... Suddenly the idea flashed across my mind that it might be iron pyrites [fools' gold]. I trembled to think of it! ... Putting one of the pieces on a hard river stone, I took another and commenced hammering it. It was soft and didn't break: it therefore must be gold but largely mixed with some other metal, and very likely silver; for pure gold, I thought, would certainly have a brighter colour.
>
> When I returned to our cabin for breakfast I showed the two pieces to my men. They were all a good deal excited, and had they not thought that the gold only existed in small quantities they would have abandoned everything and left me to finish my job alone. However, to satisfy them, I told them that as soon as we had the mill finished we could devote a week or two to gold hunting and see what we could make out of it.

Sutter's Mill, pictured as it was in 1849

The next day more gold was found and tests made, including one by the camp housekeeper, Mrs Elizabeth Wimmer, who, according to her story, had a soap-making kettle into which she put some gold and boiled it all day. When it did not tarnish, she said, Marshall was finally sure that he had struck gold.

Now he dashed to report to his partner, and the two men went to work on specimens behind locked doors, taking so long about it that suspicions were raised at the fort. Later, Sutter visited the mill and suggested that his men kept the secret for six weeks, until the work was finally finished. Though they agreed, the news simply could not be suppressed, especially when Sutter secured a three-year lease of the area around the mill from the local Indians, then tried, without success, to get the title to the land from the military governor. The messenger to the governor gave the game away to a miner who had prospected in Georgia, and who now hastened to the mill, while Sutter himself failed to keep his mouth shut.

Though the mill was finished, gold was soon being found beyond the tailrace, and on March 14 San Francisco's first newspaper, the *Californian*, carried a short report:

GOLD MINE FOUND

In the newlymade raceway of the Saw Mill erected by Captain Sutter on the American Fork, gold has been found in considerable quantities. One person brought thirty dollars worth to New Helvetia, gathered there in a short time. California, no doubt, is rich in mineral wealth; great chances here for scientific capitalists. Gold has been found in almost every part of the country.

No one paid much attention, and, indeed, there were few people to read the news, San Francisco, recently known as Yerba Buena, being a mere village of some 800 people. In all California at this key moment in her history there were some 14,000 whites, many of whom were of Spanish descent and less gold-obsessed than Anglo-Americans nearly always were. San Franciscans waited to see what their friends would do before setting out, for they feared being laughed at.

Then on May 15, as we have seen, Sam Brannan, an extrovert Mormon-turned-merchant, who was later to be a Vigilante leader, rushed through the town waving his bottle of gold dust. That did it. A stampede began and by mid-June three-quarters of the town's men had left for the mines. Meanwhile, crews deserted their ships to be followed as a rule by their captains, one of whom left his wife and daughter to look after his vessel. Both the town's papers had to close and only goods used in mining found buyers. A shovel priced at a dollar shot up to ten, while many shops were closed and on them was written: 'Gone to the Diggings.'

In this first local rush all the coastal towns were affected. Servants deserted their masters; houses were left half-built; the countryside was depopulated. One eye-witness reported how 'Every bowl, tray, warming-pan and pigin has gone to the mines. Everything in short that has a scoop in it that will hold sand and water. All the iron has been worked up into crow-bars, pick-axes and spades', while Walter Colton in Monterey, author of *Three Years in California*, memorably recalled the frantic scene in these words:

> *June 20, 1848*.... My messenger has returned with specimens of gold; he dismounted in a sea of upturned faces. As he drew forth the yellow lumps from his pockets, and passed them around among the eager crowd, the doubts, which had lingered till now, fled.... The excitement produced was intense; and many were soon busy in their hasty preparations for a departure to the mines. The family who had kept house for me caught the moving infection. Husband and wife were both packing up; the blacksmith dropped his hammer, the carpenter his plane, the mason his trowel, the farmer his sickle, the baker his loaf, and the tapster his bottle. All were off to the mines, some on carts, some on horses, and some on crutches, and one went in a litter....

Colton did not mention the soldiers stationed in California, some of them fresh from combat in the Mexican war, and all of them ill-paid, the going rate for privates being six dollars a month. Even if inflation had not been rampant in the coastal cities within a few days of Sam Brannan's electrifying appearance in San Francisco, six dollars could hardly have held soldiers to their posts, and their officers recognised it. One deserter summed up the feelings of all when he wrote: 'A frenzy seized my soul; piles of gold rose up before me at every step; thousands of slaves bowed to my beck and call; myriads of fair virgins contended for my love. In short I had a violent attack of gold fever.' So did plenty of others, for between July, 1848 and the end of 1849, 716 out of 1,290 soldiers in northern California deserted. A few were rounded up but most escaped for good, and so did the majority of the sailors in the Pacific Squadron.

Neither Sutter nor Marshall benefited from the bonanza in their midst. Marshall's was the more pathetic case, yet given that his country should have rewarded him, his defeatism and inertia in a seething period which called for action helped sour his life. He drank, sold autographed cards, drifted from place to place, then in 1872 was finally granted a $200-a-month pension by his state, which was halved the next year, then cut off altogether when he

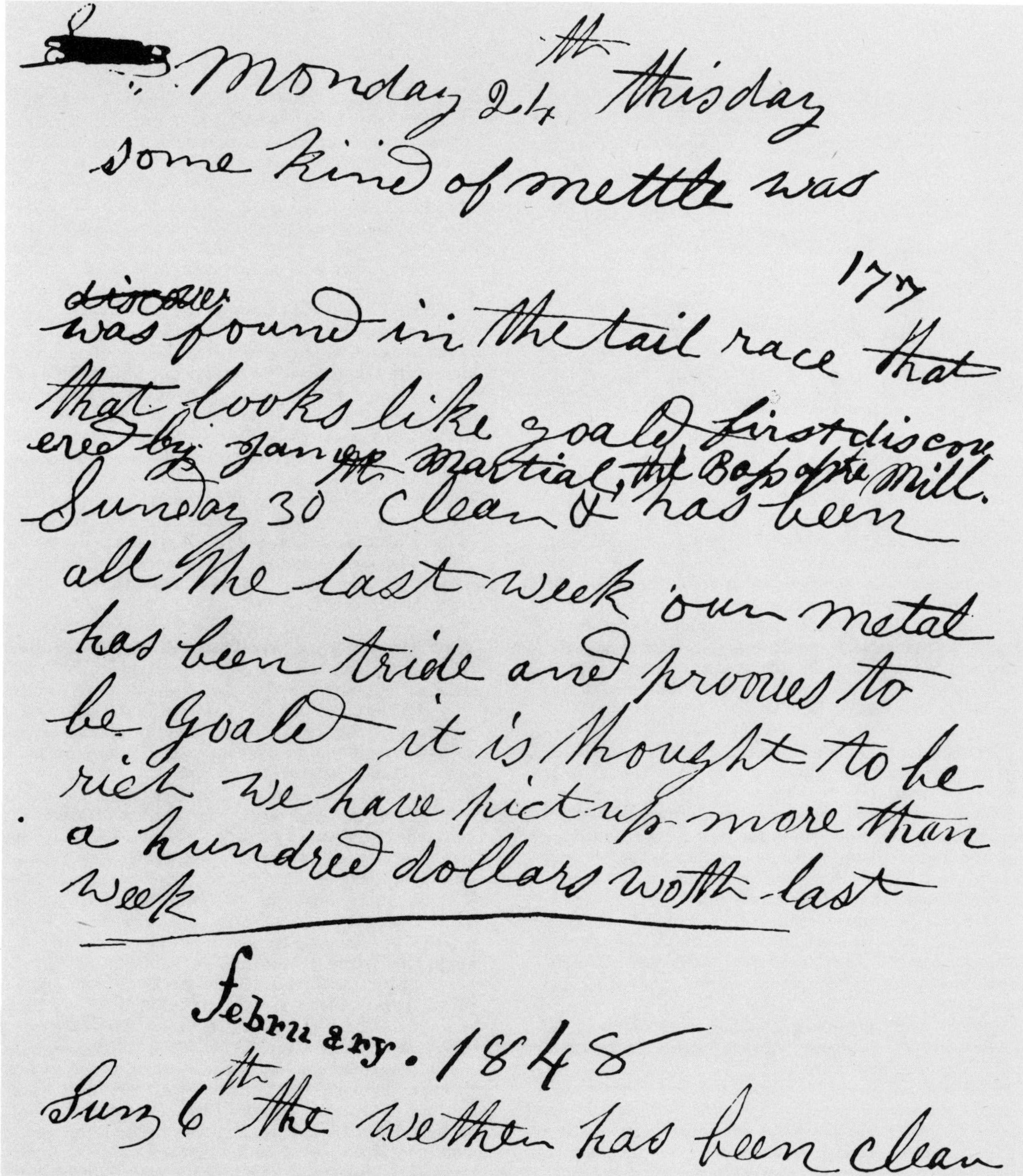
[redacted] Monday 24th this day
some kind of mettle was
discov 177
was found in the tail race that
that looks like goald first discov
ered by James Martial, the Boss of the mill.
Sunday 30 clear & has been
all the last week our metal
has been tride and prooves to
be Goald it is thought to be
rich we have pict up more than
a hundred dollars woth last
week

February. 1848
Sun 6th the wether has been clear

A diary containing the first known mention of the discovery of gold in California in 1848. The writer was Henry Bigler, a mill hand of John Sutter's

appeared drunk at the assembly chamber in the State capital, Sacramento. He died a pauper in 1885 and a ten-foot statue of him was soon erected which cost $25,000!

Sutter tried vainly to hold his empire together, though even if he had drunk less, economised more–he was an over-generous host–and been a better businessman, the task would probably have been beyond him. Finally, the U.S. Supreme Court decided that his land grant from the Mexicans was invalid. He settled modestly in Pennsylvania, dying in 1880 just when the Congress had decided that its treatment of him was unworthy. As we shall see, things were better ordered in Australia.

That first summer and autumn of the gold rush, when it was purely local, was an idyllic time. Though only some struck it rich, there was enough gold for many to find at least a modest amount. In the six months from May to November numbers in the gold-fields rose from a few hundred to about 10,000, eager prospectors rapidly heading north from Sutter's Fort as far as the Trinity River and south to the Tuolumne.

There was virtually no crime in these magical months as the prospectors rushed from place to place following every rumour. No one had yet worked out rules governing claims, no one had to compete. The shootings and lynchings would come later.

Methods of getting gold at this time were simple in the extreme. In the 'dry' diggings away from the streams gold could be picked out of crevices in the rock with a knife. Edward Buffum, quoted at the start of this chapter, was describing a success at a dry digging in Weaver's Creek. He was a New Yorker who had been sent to California as a soldier before the rush, to be mustered out in time to take part in it. One day in October, 1848 he found the right crevice after a long search, which 'appeared to be filled with a hard bluish clay, which I took out with my knife, and there at the bottom, strewn along the whole length of the rock, was bright, yellow gold, in little pieces about the size and shape of a grain of barley.' No wonder he exclaimed: 'Eureka!'

The 'wet' diggings were on the banks of streams, where 'placer' gold was procured by washing away the sand and gravel. There were two widely used methods in 1848, the first being panning, the basic, if backbreaking, way to obtain gold. To squat in icy water under a blazing sun was never less than arduous, and it took up to twelve minutes to wash a pan. The miner would fill his iron or tin pan–which could also be used for washing or frying–with dirt, then swirl it round under water. Then he would raise it out, keep swirling it round, while jerking it in and out of the water. The lighter dirt was thus washed away leaving the gold, which would sink to the bottom because of its higher specific gravity.

Fortunately, the rocker, or cradle, was introduced at the very beginning of the gold rush, possibly by a miner who had learnt to use it in Georgia. Here is one account of it, along with a vivid picture of the scene at a point on the American River on August 17, 1848. The writer is Colonel Mason, the Military Governor of California:

> The hill sides were thickly strewn with canvas tents and bush arbours; a store was erected, and several boarding shanties in operation. The day was intensely hot, yet about two hundred men were at work in the full glare of the sun, washing for gold–some with tin pans, some with close-woven Indian baskets, but the greater part had a rude machine, known as the cradle.
>
> This is on rockers, six or eight feet long, open at the foot, and at its head has a coarse grate or sieve; the bottom is rounded, with small cleets nailed across. Four men are required to work this machine;–one digs the ground in the bank close by the stream; another carries it to the cradle and empties it onto the grate; a third gives a violent rocking motion to the machine; whilst a fourth dashes on water from the stream itself. The sieve keeps the coarse stones from entering the cradle, the current of water washes off the earthy matter, and the gravel is gradually carried out at the foot of the machine, leaving the gold mixed with a heavy fine black sand above the first cleets.
>
> The sand and gold mixed together are then drawn off through augur holes into a pan below, are dried in the sun, and afterwards separated by blowing off the sand. A party of four men thus employed, at the lower mines, averaged $100 a day. The Indians, and those who have nothing but pans or willow baskets, gradually wash out

Sutter's Fort in 1848, where James Marshall made his great find. He died poor, but was given a monument – four years after his death.

As far away as Britain gold fever raged, as this 1849 cartoon indicates

A REGULAR GOLD DUSTMAN.

"Hollo! Where are you off to now?"

"Oh! I aint a going to stop here, looking for Teaspoons in Cinders. I 'm off to Kallifornier, vere there 's heaps o' Gold Dust to be had for the Sweepin'."

Time for lunch beside a long tom near Auburn, California, in 1852

the earth, and separate the gravel by hand, leaving nothing but the gold mixed with sand, which is separated in the manner before described. The gold in the lower mines is in fine bright scales, of which I send several specimens.

For the record, the usual word for the cleets that held the gold is 'riffles', and gold was found in three basic forms: nuggets, dust and flakes. The going rate for it varied. In 1848, it was sometimes as much as $10 an ounce, in 1849, $16.

Though most of the gold was sought by small groups of miners, some employed Indians, who, until they realised what was going on, were grossly exploited. Later, as we shall see, they were more savagely treated.

The system of claims, when it existed at all, was simple: miners recognised a claimant who had cleared the top soil from any portion of the bar of a river. Most men went armed, but it can hardly have been that which allowed bags of gold to lie about unguarded in that magical period.

Of course, not all was magic, for already some were suffering from rheumatism, scurvy, dysentery and total exhaustion, but for many until the heavy rains and cold of winter set in, the places where they sought gold seemed like a wonderland.

By the end of the year $10 million may have been extracted from the dry and wet diggings of the new Eldorado, and by then the news had reached the outside world. Gold-carrying ships brought it first to South America, Mexico and Hawaii before it reached the eastern United States. The stage was set for the arrival of the latter-day argonauts, the Forty-Niners.

Cartoonists had a field day with the Forty-Niners. Here is Mr. Golightly heading for California in 1849

Panning for gold near Virginia City, Montana, in 1871

403

The Forty-Niners

Oh! California!
That's the land for me,
I'm going to Sacramento,
With my washbowl on my knee!

'The farmers have thrown aside their ploughs, the lawyers their briefs, the doctors their pills, the priests their prayer books, and all are now digging for gold,' wrote a Californian to the *Philadelphia North American and United States Gazette*, which published the letter on September 14, 1848, together with the writer's bold statement that ten thousand men in ten years could not exhaust the supply. It certainly caused more stir than reports of Marshall's discovery had made in New York and elsewhere in the East back in August, but the real turning point was President Polk's message to Congress on December 5th.

Polk was acting on information received from Colonel Mason, whose account of the cradle we have already seen. Appalling lies and exaggerations were later to stem from gold-mining areas, made grosser in the repeating, but Mason's was heady fact. Said Polk: 'The accounts of the abundance of gold in that territory are of such an extraordinary character as would scarcely command belief were they not corroborated by the authentic reports of officers in the public service, who have visited the mineral district.' And he mentioned that the mineral wealth in the area drained by the Sacramento and San Joaquin river system would pay the cost of the war with Mexico a hundred times over.

Only two days later, Lieutenant Loeser, Mason's courier, reached Washington with a tea caddy full of pure gold samples, weighing in all 230 ounces, 15 pennyweights, and 9 grains. America promptly succumbed to gold fever, even though most people knew as much about California as they did about the far side of the Moon. The wildest rumours were circulated and some thirty guidebooks were rapidly published, most of them totally useless.

Preachers, lecturers, journalists, clairvoyants and advertising copywriters all contributed to the frenzy. Horace Greeley of the *New York Daily Tribune*, who was later to immortalise someone else's order '*Go West, young man*', exulted that 'fortune lies abroad upon the surface of the earth as plentiful as the mud in our streets. . . . The only machinery necessary in the new Gold mines of California is a stout pair of arms, a shovel and a tin pan.' He told his enflamed readers that even without the pan they could hope to make between 50 and 60 dollars a day and still have plenty of leisure.

Naturally it was the young who were most affected by the frenzy, though Hollywood has implanted the idea that the typical miner was 50 or more with a beard bigger than his face. But young or old still had the same basic problem–how to get to Eldorado.

There were a number of alternatives. Many Easterners sailed around the Horn, a tedious, expensive trip which on average took 168 days, and which softened a man for the rigours to come (though, as we shall see, the journey had its own rigours!). Others took the short cut across the Isthmus of Panama long before there was a canal, when

Later travellers enjoyed comfier voyages on more modern vessels, even though most of the gold in California had already been found

The mining area of the American West and British Columbia

BRITISH COLUMBIA
Quesnel
Thomson R.
Fraser R.
Victoria
WASHINGTON
Columbia R.
Umatilla
Portland
OREGON
Boise
IDAHO
Salmon R.
Snake R.
MONTANA
Missouri R.
Virginia City
NORTH DAKOTA
SOUTH DAKOTA
Deadwood
Lead
WYOMING
Sweetwater
Fort Laramie
South Pass
NEBRASKA
Platte R.
Humboldt R.
Sacramento R.
Virginia City
Carson City
American R.
NEVADA
Sacramento
Aurora
San Francisco
S. Joaquin R.
Salt Lake City
UTAH
Colorado R.
Denver
Leadville
KANSAS
Arkansas R.
COLORADO
CALIFORNIA
Los Angeles
ARIZONA
NEW MEXICO
Santa Fe

the chances of avoiding killer diseases were less than evens, and when sick and fit had to face extortionate price rackets. The same held good of the less-used Nicaragua route.

Some braved the bandits of northern Mexico, and many Southerners made the long journey through the Southwest and on up to California. But the classic route was straight across the continent from the Missouri, a 2,000-mile trip which at first followed the Oregon Trail, then the already-blazed California Trail.

Britons and Europeans sailed via the Horn, while many Australians, including ex-convicts who were to cause much trouble in San Francisco, headed across the Pacific along with Hawaiians and thousands of Chinese. Few nations were to be unrepresented on the gold-fields.

By the end of 1849 the population of California, excluding Indians, had risen from around 26,000 to 115,000. The word to describe the adventurers was Argonauts, though they are remembered by the later name of 'Forty-Niners'.

The sea-voyage from the East was around 13,000 miles, and ships began leaving as soon as President Polk had made his historic statement. More than fifty set out from New York alone in February 1849, and by the summer the American flag had almost disappeared from the harbours of the world, nearly every vessel being California bound, including cargo boats and the whaling fleet.

Many hastily refurbished ships were old and slow, and some were unseaworthy. There were a few crooked shipping agents who made a nice profit out of selling the same cabin to a number of would-be passengers, and sometimes food was rotten before the voyage had even started. 'There were two bugs for every bean' in one ship. Easterners especially formed companies for the trip, sharing the cost of tickets and equipment and planning to share the proceeds the other end, though in the event nearly all these associations disbanded in California when it became apparent that mining was–in the early days–for individuals or small groups.

The companies sometimes numbered several hundred and were organised for several years, complete with leaders. Even in the class-conscious East democracy was sparked off by the prospect of gold, one Massachusetts doctor joining a party going West, which was led by his own coachman.

The voyage round the Horn, especially in the Southern winter, could be rough in the extreme, and only the fast freight-carrying clippers could hope to do the journey in 100 days, but at least the sea Argonauts could take plenty of baggage along. Many took too much, including Enos Christman who travelled with 18 new checked shirts, 17 pairs of new heavy pantaloons, 12 flannel shirts, also new, 7 coats and 5 waistcoats, most of which he had to give away or get rid of on arrival. Nearly all took weapons and, fortunately for us, an extraordinary number decided to keep diaries.

The sea-goers found it harder and harder to fill their journals as day succeeded day and monotony and sea-sickness took their toll. Occasional landfalls at Rio and Callao brightened things and allowed passengers a change from foul-tasting water, weevilly biscuits and other routine hardships. Meanwhile, there were books to read for some, card-games, concerts and other diversions, including cursing the captain and the crew until the Horn was reached when, in that fearful region, every Argonaut soberly realised that his survival was in the sailors' hands.

Some ships at least did things in style. The *Rising Sun* celebrated the Fourth of July with bugles, a reading of the Declaration of Independence, military music and a dinner of roast goose, plum pudding, mince pies, figs and nuts. And no doubt some pious New Englanders at least were

THROUGH TO SAN FRANCISCO
Atlantic & Pacific Transportation Line
For the Conveyance of Passengers and Freight
SAN FRANCISCO, AUSTRALIA,
NEW-ORLEANS.
The 5th and 20th of every month.
WILLIAM MAYNARD YOUNG & CO.,

The way to California via Cape Horn was the most popular route with Easterners, slow as it was compared with more hazardous journeys across the Plains, Mexico or the Isthmus of Panama

eager to do good works as well as make their fortunes. The President of Harvard exhorted the travellers aboard a ship named in his honour, the *Edward Everett*: 'Take your Bible in one hand and your New England civilisation in the other and make your mark on that country.'

The Argonauts were driven to practical jokes to keep their spirits up, and some stole from the galley, others debated, struck up friendships which were rarely lasting, and, as the long voyage went on, tried to get a little privacy. More than half the voyagers belonged to a company, which, apart from sharing costs, as already noted, allowed for shared misfortunes as well, for the members usually came from the same area. Along with the gold-seekers on the Horn route came a number of businessmen, gamblers, aspiring politicians and 'soiled doves', though the old rhyme states that 'The miners came in '49, the whores in '51'. Whatever the truth of that, by 1851 respectable women were very much in evidence in California.

The legitimate fare via the Horn was $500 dollars, which today would be the equivalent of more than $4,000 (£2,350), but some were charged twice the correct amount. An unknown number of ships, probably not many, sank, usually in the Horn area or braving the Straits of Magellan, an even more hazardous route, but none are known to have turned back. What is certain is that few captains could hold their crews at the height of the Gold Rush. J. B. Frémont recalled the scene in San Francisco and its harbour in 1849; 'A few low houses, and many tents, such as they were, covered the base of some of the wind-swept treeless hills, over which the June fog rolled its chilling mist. Deserted ships of all sorts were swinging with the tide.'

Though the Nicaragua route to California became quite popular in the '50s, Panama was the classic short cut, attracting over 6,000 daring travellers in 1849 and a record 24,000 in 1852.

The United States already had an arrangement whereby its citizens had freedom of transit in New Granada (of which Panama was then part) in return for American recognition, and, luckily, just as the gold rush was beginning, a mail service across the Isthmus had been arranged. The U.S. Mail Steamship Co. was to run the service from New York to the town of Chagres on the east coast of New Granada, while the Pacific Mail Steamship Co. would provide steamships to San Francisco, then sailing ships further up the coast. But between the two lines lay a region ridden with cholera, dysentery, yellow fever, wild life not to every taste, jungles, swamps and ticket scalpers, a 70-mile journey full of interest, which could be achieved in five days, given a lot of luck.

The crossing began with a 50-mile trip up the Chagres River in dugout canoes known as bongos, with overnight stops at native villages. The boatmen were mostly of mixed Indian, Spanish and Negro blood, and if treated decently by the Argonaut would do their jobs reasonably well as long as they were given regular draughts of liquor and adequate rest periods. Three or four would pole the 25-foot long hollowed-out logs, carrying up to four passengers plus baggage.

Not surprisingly, prices rapidly spiralled from $10 to $50 for the trip, even though singing was provided free in the form of *Oh Susanna* and *Yankee Doodle*, learnt parrot-fashion. But at least one traveller, a Dr McCollum from New York, was honest enough to admit that if genuine Yankees had the running of the service, 'there would be such a fleecing and extortion as the Isthmus has not yet witnessed.'

When Gorgona or Cruces were reached the Argonaut took to his feet (assuming he was still fit to travel), or went by mule. The climate was better for the rest of the journey, and sometimes natives could be found who would carry the weary travellers on their backs. One wrote: 'My Indian made me up into a bundle of the easiest form to himself and threw me on his back as a porter does without troubling about my own painful and fatiguing position.' The discomfort was more than partly due to his having to sit on a plank called a tabillo. The bongo and the mule were both more popular.

The first Argonauts met with bitter frustration at Panama City, which was soon swarming with people, California bound, all fearful that the gold would be gone before they arrived. The bottleneck started at the very outset, when the *California*, destined to ply up and down between Panama and San Francisco, reached the former from New York via the Straits of Magellan. That was on January 17, 1849, by which time well over 1,000 frantic Argonauts were waiting in Panama City. As the *California* was built to take 250 passengers, and as she had taken aboard about 70 in Peru, there were ugly scenes in the city, but finally she sailed on February 1 with some 365 aboard, some of whom are said to have paid $1,000 for room in the steerage.

She reached San Francisco on February 28, and the wretched Captain Cleveland promptly lost all his crew, whom he needed to return to Panama for the next load of Argonauts. Though only paid $250 a month himself, he was forced to hire an engineer and a cook at twice that amount (each), firemen at his own salary and ordinary seamen at $200 apiece. Later captains sometimes clapped their crews in chains on reaching San Francisco.

Back in Panama City there were liable to be waits of up to four months before a space aboard a ship could be found. It was as well that the Panama route tended to be used by the fairly well-to-do, for hotel rooms cost about $8 a week, assuming that a room could be got. The old city soon sprawled an insanitary shanty-town, while the miserable Argonauts took to the bottle, provoked riots by entering the cathedral drunk and in their hats, and went down with various fevers. Things gradually improved–they had to–and in 1855 a railroad spanned the Isthmus, long before which the shipping situation had improved and so ceased reducing would-be gold-miners to agonies of frustration.

William Kelly reached St Louis in the early Spring of 1849:

> . . . the further west I proceeded, the more intense became the Californian fever. California met you here at every turn, every corner, every dead wall; every post and pillar was labelled with Californian placards. The shops seemed to contain nothing but articles for California. As you proceeded along the flagways, you required great circumspection, lest your coat-tails should be whisked into some of the multifarious Californian gold-washing machines, kept in perpetual motion by little ebony cherubs. . . .

Kelly, an Irish journalist, was one of those who was poised to make the most popular crossing of all, and the fastest, despite its dangers. More than 20,000 followed the emigrant trail across the plains and the Rockies in 1849. That spring saw thousands of restless men and women camped along 200 miles of the Missouri between Kanesville (now Council Buffs, Iowa) and Westport Landing (now Kansas City, Missouri). Booming Independence, Missouri, which had been the main starting point of the Oregon Trail in the early '40s, had lost its supremacy because the temperamental Missouri River had changed course enough to deprive her of her steamboat landing place.

The main supply centre now was St Joseph, and between them all these small frontier communities could produce enough wagons, mining gear, food and animals. The Forty-Niners contained a higher percentage of greenhorns than the earlier and more stable would-be settlers in the far West, who were mostly of pioneer stock, and for all the newcomers forming themselves into companies (as had earlier emigrants) they were less amenable to the discipline of the trail, some regarding obligations 'about as much as singing psalms to a head horse'. But at least they could draw on the experience of earlier travellers.

Oxen and mules were the most favoured animals for the 2,000-mile trip, and some had both. The former, slow, strong and reliable, could manage 12 to 15 miles a day, and enjoyed food that horses and mules would not touch. Mules required more care, but could go up to forty miles a day in flat country, and were certainly the answer to the mountains, so secure were their footholds. The animals were usually bought unbroken, which resulted in some very lively hours for those unused to their ways.

Then there was the all-important wagon to be bought. Many were strengthened farm wagons, the wise newcomer having a small one of seasoned hardwood, with three yoke of oxen. The canvas that was stretched over the bows could be weatherproofed with beeswax and linseed oil, and when, as often happened, the wheels were painted red and the body blue, the result was distinctly patriotic. The name or the initials of a company were painted on the canvas, or, more simply, the owner's home town, or some permutation of words like 'Gold and California'.

A few did without wagons, using hand-carts or carrying their baggage on their backs. One man is known to have pushed a wheelbarrow across the plains, and another footslogger took a cow, which carried his belongings and provided him with milk.

Generally, far too much was taken, and later abandoned. Weapons, tools and food were the essentials, with a handful of possessions and basic clothing, but when the going got steadily tougher the trail became littered with abandoned stoves, chests-of-drawers, mining equipment, tables and other luxuries, as some things really were: harps, four-posters, pianos.

It was vital to reach the Sierra Nevada, the last barrier before the goldfields, early enough to get through the passes before snow blocked them, but too early a start was useless. The young grass had to be grown enough for the livestock, and the winter floods had to have gone down, leaving rivers within their banks. So the normal starting time was in May. By May 18 of 1849, 2,850 wagons had crossed the Missouri at St Joseph and 1,500 elsewhere.

So the wagon trains set out for Eldorado, and nearly all of them, large or small, were formed into a semi-military organisation with a captain, quartermaster and lieutenants, all of them elected and all liable to be changed en route if things did not go well. The captain was expected to be a superman without legal authority, in charge of everything from the starting time each day to the selection of the next campsite and how it was to be defended.

A striking glimpse of Custer's expedition in 1874 into the Black Hills of Dakota, allegedly in the interests of science

Wagons took their turn in the column, those at the head one day being at the rear the next and working their way up again. Accidents were frequent, especially as the wagons had no brakes, while rivers that could not be forded were crossed by raft, or the wagons were floated across.

The first few weeks on the trail were almost a picnic, with plenty of rich grass, water and wood, spring flowers everywhere, still some farms to pass, and a wide sky above, but gradually the going got tougher. Each night the traditional circle of wagons was made, linked together in case of attack, but at this time there was little or no trouble from Indians beyond loss of livestock.

Sometimes the Indians tried to collect a dollar a head from the Argonauts on the reasonable assumption that their land was being crossed. One California-bound train's reaction was later recalled by Sarah Royce, who was with her husband and two-year-old daughter:

> The Indians were then plainly informed (by the Captain) that the company meant to proceed at once without paying a dollar. That if unmolested, they would not harm anything; but if the Indians attempted to stop them, they would open fire with all their rifles and revolvers. At the Captain's word of command all the men of the company then armed themselves with every weapon to be found in their wagons. Revolvers, knives, hatchets, glittered in their belts; rifles and guns bristled on their shoulders. The drivers raised aloft their whips, the rousing words 'GO 'long Buck'–'Bright!' 'Dan!' were given all along the line, and we were at once moving between long but not very compact rows of half-naked redskins. . . .

In fact, many Indians helped the Forty-Niners. There were still a few more years of peace on the plains between red men and white.

The chief killer on the journey was cholera, which was particularly widespread and terrifying in 1849. It struck with dreadful suddenness and the less responsible wagon trains simply abandoned the sick, often leaving them completely alone.

Less perilous than the cholera, but unpleasant enough, were two constant companions of the trail, diarrhoea and nappy rash, and not until Fort Laramie in Wyoming could wives enjoy a real washing day, while stores were replenished and wagons were mended. Cooking was mainly done in Dutch ovens, those without a woman cook usually confining themselves to bacon ('sowbelly'), hot bread ('biscuits') and plenty of coffee. Buffalo in their millions still roamed the Plains giving excellent meat, and their dung, or 'chips', provided fuel on the treeless stretches of the lone trek. Buffalo meat could be jerked (cured) and

used as reserve stock. Hunting the buffalo turned out to be harder than the Forty-Niners expected, and there were often casualties when wounded beasts charged their assailants and their horses.

Where did the great adventure cease to be fun? Perhaps in what is now mid-Nebraska, where the Platte divides into North and South, when the timber had long given out, and where upright planks with simple inscriptions proclaimed newly dug graves. Now badly-driven cattle were going lame, now the debris of earlier wagon trains began to litter the ground. As well as the rivers to be crossed were great landmarks rising out of the plain, Court House Rock, Jail Rock, Chimney Rock, Scotts Bluff. At the last, and near Fort Laramie, you can still see the ruts made by the wagons of the Forty-Niners, bitten deep into the rocky soil.

Past the famous fort and on towards Fort Bridger, 390 miles away and in the mountains, the travellers were now in tatters, their faces burnt by both sun and alkali. Horses died at poisonous water holes. One of the songs the wagon-trainers sang had these two lines:

If I knew once what I know now
I'd have gone around the Horn.

But then came the Sweetwater, aptly named, and Independence Rock, covered with the signatures of those who had passed by since the first crossings to Oregon.

They went through South Pass, not much scenically, but a place which divided the Atlantic and Pacific watersheds. Past the Continental Divide, there were by now several routes and 'cut-offs', none of them less than exhausing, to rejoin the main California Trail for the miserable 300-mile journey along the Humboldt River. Steadily it got more putrid–it was also known as the Hellpot and Humbug–and now the route was truly littered with cast-off belongings and broken-down wagons, as the river turned yellowy green and became alkaline enough to kill cattle that drank it. There were suicides along the Humboldt.

A choice of nightmares now faced the Forty-Niners. The usual choice in '49 was the day-long crossing of the Humboldt Sink, which had no drinkable water, and beyond it was Destruction Valley. Only by night was it just endurable, for those animals that survived. 'Expect to find the *worst desert* you ever saw and then find it worse than you expected. Take water, *be sure* to *take enough*,' wrote one migrant to those following, who promptly filled even their gum boots and a rubberised blanket with water.

Blessed relief came at the Carson River, where it was normal practice to send back relief to those staggering towards it.

Now came the last barrier, the mighty fortress of the Sierra Nevada range. It would be foolish to suggest that the nightmare was over, for many wagons that had survived so far finally broke down. Those mules which had not died, or did not now vanish over cliffs, became invaluable pack animals. So did the men. But there was water and

A wagon train crossing the Smokey Hill River in 1867.

The forest of masts in San Francisco harbour in 1851. Only two years before it had been a sleepy village

food for beasts and humans and the worst was over–if they were through the mountains by early September. Latecomers were likely to be trapped and some died in the snows. One party, even later than the rest, headed south and accidentally became trapped in the wastes of Death Valley, that 'seventy-five mile strip of perdition' 110 feet below sea level.

There were 13 men, 3 women and 6 children in the lost party, and finally, two of the youngest and strongest men went on for help. They reached a mission and returned with food and horses to find the survivors huddled under the wagons more dead than alive. One of the rescuers, William Manly, fired a shot and a man appeared. 'Then,' wrote Manly, 'he threw up his arms high over his head and shouted–"The boys have come! The boys have come."... Bennett and Arcane caught us in their arms and embraced us with all their strength, and Mrs Bennett when she came, fell down on her knees and clung to me like a maniac, in the great emotion that came to her, and not a word was spoken.'

It took this party more than a year to reach the gold-fields instead of the regulation three months plus, but their story is one of loyalty and courage, and the former was not always in evidence during the great crossings. Meanwhile most of the Forty-Niners had reached their journey's end in the autumn. We shall see the way they lived in the next chapter. Whether they had come by sea, across the Isthmus, through Nicaragua, Mexico, through Arizona, or across the Plains and the Rockies, they were the survivors. Back on the trail the graves remained. J. Goldsborough Bruff, an ex-West Pointer who crossed with the Washington City and California Mining Association as its leader, somehow finding time to take notes and draw scenes from the journey of 120 days, noted some of the graves between Chimney Rock and Green River. Three can stand for all the Forty-Niners who never made it:

Jno. Hoover, died June 18.49
Aged 12 years. Rest in peace,
Sweet boy, for thy travels are over

Jno. Campbell,
of Layfayette Co. Mo.
came to his death by the
accidental discharge
of his gun, while riding
with a friend, June 21
1849, Aged 18 years.

Robert Gilmore
and wife,
Died of cholera,
July 18th, 1849.

▶28 A view of a mining area by an unknown artist. Miners learnt to use the cradle (or rocker), the long tom and its improved version, the sluice, yet only the humble – and back-breaking – pan ensured that the final particles of gold did not elude them

Seeing the Elephant

It had confidingly come to our ears that someone had affirmed that he had seen a man who had heard another man say that he knew a fellow who was dead sure that he knew another fellow who, he was certain, belonged to a party that was shovelling up the big chunks.

D. R. Leeper: *The Argonauts of 'Forty-Nine.*

At spots like Skunk Gulch, Shinbone Peak, Ground Hog Glory and Lazy Man's Canyon they sought gold and some of them found it. The whole amazing experience was summed up in the phrase, 'seeing the elephant', which is odd enough to bear explaining.

Once upon a time there was a farmer who had never seen an elephant and dearly longed to do so. One day a circus came to town, so he filled his wagon with vegetables and eggs and headed for market, meeting the circus parade on the way. There at the head of it was the longed-for elephant.

The farmer was in heaven, but his horses were not, rearing up, overturning the wagon and running away. Eggs and vegetables were thrown along the wayside, but the farmer didn't care. 'I have seen the elephant,' he exulted.

Technically, 'seeing the elephant' only applies to the Californian Gold Rush, with its occasional joys and fantastic finds and its far more frequent disappointments, but it could really apply to all the gold rushes of the 19th century. It sums up the exaltation of having reached the longed-for spot, the feeling a traveller has when at last he visits somewhere he has wanted to see all his life and can console himself forever with the thought: 'I've been there.'

Sutter's Fort was the magnet for the Forty-Niners coming overland, and for many who landed at San Francisco, that greatest of all cities spawned by the gold rushes, to which we shall return. Upon Sutter's land hordes of gold-seekers squatted and worked, and the town of Sacramento shot from nothing in November, 1848 to a booming centre of 12,000 people by the end of 1849.

Finds in the northern mining area around Sacramento and the southern, based on Tuleburg, later called Stockton, varied vastly. One man took $16,000 worth of gold from along the Feather River in only eight working days; three men uprooted a tree and found $5,000 worth below it; some made even bigger finds, but the majority may have averaged about $100 a month in 1849 and $50 by 1852. If that seems reasonable, it should be remembered that goods needed by the miners were wildly expensive, and, as we shall see, those who sold them were the most likely to become rich. Just occasionally absurd pieces of luck occurred, as when a miner found two and a half pounds of gold in fifteen minutes at a time when it was selling for $16 an ounce. As for prices, an egg could cost up to $5 in a restaurant, a loaf which would sell at 5 cents in the East, could cost up to 75 cents in San Francisco. And getting to the mining area from the coast was slow and expensive until a steamer service up the rivers began late in 1849, while the trails beyond the rivers were difficult going for Argonauts whose muscles, if any, had been slackened by months at sea.

Many Argonauts wore red flannel shirts, many more blue or grey flannel or red calico, and heavy woollen trousers and strong boots completed the basic outfit, along with generally broad-brimmed, low-crowned felt hats. Standards of hygiene were low, though Sunday was officially washing day. It took a considerable effort to avoid becoming lousy.

The men's diet was chiefly jerked beef or salt pork, plus bread, and decent cooks were rare, especially as gold was always the priority. Some starved to death and many went down with scurvy because of lack of vegetables. In the first winter of the gold rush, 1848-49, it was rampant.

Later, cholera, typhoid, meningitis and rheumatic fever were to scourge both towns and mining areas, and many were weakened for life. Not until the late '50s was there a reasonable number of good doctors.

With gold fever at its height, and involving frequent movement, housing was low on the list of essentials, only camps which seemed likely to survive having true houses as opposed to shacks and tents. The unwise suffered horribly in the winters.

The miners soon improved on the pan and the cradle, though these remained the staple tools of the trade. The 'long tom' took the cradle principle a stage further, with a 12-foot-long trough running down to a perforated iron 'riddle', below which was a riffle box; and the sluice went even further, being a series of riffle boxes down which water flowed, as miners shovelled in the dirt. The importance of water became paramount, especially when claims were staked far from it, and companies were organised to build canals and ditches, also flumes, which were open ditches normally made of heavy boards, and, later, of iron pipe. By the mid-50s it was inevitable that mechanised mining, complete with earth-rending machines, should be more important than the individual gold-seeker, who by then was heading elsewhere. And by then hydraulicing had become a destructive feature of the scene, with great jets of water blasting whole hills away to get at gold, while filling valleys with gravel, sand, rocks and mud.

Despite these advances and portents of things to come, only the humble pan, if properly handled, could be sure to keep the last particle of gold from vanishing, and the symbol of this and later gold rushes was the lone prospector, with his pan and his overloaded mule, wandering from place to place, restless for gold.

For all the disappointments, the magic moments kept occurring. Those who saw the musical *Paint Your Wagon*

The Elephant's message is clear enough, while below it is that notorious mining town, Tombstone, Arizona

SEVEN-LEAGUE-BOOTS EXPRESS!
DIRECT TO NEVADA!
THE LAND OF
ROCKS OF SILVER,
Streaks of Gold,
MOUNTAINS OF COPPER,
MINES OF SULPHUR.
Hie! ye Penniless, Hie!
To Plautus's Treasury.

may have imagined that the burial scene was fiction, but it really happened near Carson Creek. A miner having died, his friends decided to give him a good send off, and one of them who had been a preacher before the lure of gold called him from the South, did the honours at the graveside while the rest of the dead man's friends knelt alongside it. There was no stopping the preacher, who presumably was somewhat short of opportunities in those parts, and the congregation's attention began to flag. One or two began fidgeting with the loose earth, running it through their fingers, when one of them suddenly roared: 'Colour!'

There was no doubt about it: there were distinct signs of gold.

The preacher shouted- 'Congregation dismissed!', the body was rapidly raised from what should have been its final resting place, and everyone, the preacher included, began to dig, to great effect.

More than a quarter of the estimated 85,000 men who made up the Forty-Niners in California were not American citizens, and though there is plenty on record to suggest that the majority of miners were generous and large-hearted, racial intolerance, especially against Mexicans and Chinese, and, above all, against Indians, grew steadily. Despite the fact that California had only just become American, suddenly chauvinist citizens began to hate the idea of foreigners getting away with gold. Britons and Australians (murderous ex-convicts excepted) probably came off best, and other Europeans were tolerated, but a Foreign Miners' Tax was passed in 1850, demanding 20 dollars a month, though this was later lowered and, in 1853, became four dollars a year. In fact, it was aimed chiefly at the Mexicans and the Chinese, the latter by 1850 constituting almost a fifth of all the miners.

The Chinese were loathed and/or despised, partly because they worked so hard and lived so simply. The least of their troubles was being on the receiving end of practical jokes, one being the removal of a pigtail to 'make a good Californian' out of a man. Their liking for opium rather than alcohol was deeply resented, and they were often simply driven away from their diggings. They joined not very sinister tongs, which were part rotary club, part accommodation and job agencies, though occasionally more lurid events stemmed from the organisations, most notably at Weaverville in 1854.

There, the town's two tongs suddenly went to war in front of 2,000 excited non-Orientals, for what reason nobody ever knew–except the inscrutable Chinese. The large audience was due to the fact that the local smiths had been asked to make long pikes, and equally long pronged spears and lethal swords, and they passed the glad tidings on. Despite numerical inferiority, one tong a mere 150 strong, beat the other, the Canton City Company numbering 400, by better tactics.

The hostility to the Mexicans and Spanish Americans tended to be worst in the north where they were fewer in number and could be driven off their claims. The fact that many were more skilled at finding gold than the average Forty-Niner did not help, and that unattractive nickname 'greaser', which was applied to Mexicans and South Americans alike, showed a racial contempt which has lasted in some quarters to this day.

As for the story of the Indians, it is the true blot on the basically exhilarating picture of this first and greatest gold rush. Welcomed as workers in 1848, they were driven off the gold-fields in 1849 and hostility soon turned to enmity, with the inevitable result of atrocities on each side leading to more atrocities. But the Indians were not the fighting warriors of the Plains, being mainly unformidable opponents. The story is a sorry one of massacres of entire bands, gang rapes, prostitution, and thousands of children kidnapped between 1852 and 1867 to become near slaves. As William Brandon has written in *The American Heritage Book of Indians*: 'It was not good to be a California Indian in the 1850s.'

The heart of the mining area in the early years of the gold rush was known as the Mother Lode, the lode being long, uniform, thick and near other veins of gold-containing quartz. Of varying width, it stretched some 70-120 miles, depending on which authority defined it.

In the Mother Lode country the many mining camps were usually known as 'towns', even if some of them rapidly became ghost towns when the local gold ran out. Centre of the town was the store, even if at first it was little more than a shack or even a tent, and the storekeeper, ex-miner or not, good or bad, was the key member of the community.

'Store' is perhaps a misnomer, for a booming camp demanded and got something more. Soon it became a saloon, hotel and social centre as well. Important as the owner was, not least in the matter of credit, his accommodation tended to be primitive in the smaller camps, though better than the average miner had become used to beside his claim.

Food tended to be basic at the best, unless there was an Italian, French or German cook on the premises, but there was plenty of drink even if it was typically Frontier. It was often paid for in gold. Those who were prepared to pay for it could get choice liquor, the rest put up with rot-gut whiskey: coffin varnish, stagger soup, tornado juice and the fabulous Forty Rod and Sixty Rod, whose mere smell could apparently knock a man down at 40 paces, even if he was round a corner!

Gambling was hugely popular, and, as always, the well-

dressed, well-controlled, temperate professionals came off best. As we shall see, San Francisco had high class gambling establishments, but even the smaller towns sometimes ran to real gambling houses, as opposed to a corner of the store.

And then there were the women, not many of them, but a growing band, even though in 1850 less than 8% of Californians were female. All-male dances were only too frequent in the early days as the lonely miners let off some of their high spirits, and the appearance of a woman, however homely, tended to attract a crowd of respectful, awed onlookers. When Mrs Galloway arrived at Downieville, the miners–who had been totally without female company–rushed out to greet her and carried her and her mule into town. When Herman Reinhart reached Humbug City in northern California in 1852 he found a saloon kept by a man named Nobles, whose 'Fancy Woman' was a fine-looking German girl who sold drinks at 50 cents. Many miners paid the 50 cents just to see her, and some said they had not seen a woman for five years.

Gradually enough prostitutes reached California to satisfy the immediate needs of at least some of the miners, but these 'soiled doves' were not what many more ultimately wanted. It was a sentimental age. Mother and sister back home were greatly missed and so was the love of a good woman.

The burial service was going well until someone spotted 'colour', at which point avarice took over from holiness

Worthy Eastern ladies tried to remedy the lack of decent women, notably an ex-matron of Sing Sing, Mrs Farnham, who hoped to bring over 100 clergyman-vouched ladies to the West, but when she embarked with only three, she kept quarrelling with the captain and was finally turned out of the ship in Chile. A Miss Pellet hoped to attract 5,000 'virtuous New England women' to the gold-fields, but she never even got started.

Yet some of the Miners longed for wives, and within reason would accept anyone in skirts. One lady advertised her charms:

A HUSBAND WANTED

By a lady who can wash, cook, scour, sew, milk, spin, weave, hoe (can't plow), cut wood, make fires, feed the pigs, raise chickens, rock the cradle (gold rocker, I thank you, Sir!), saw a plank, drive nails, etc. These are a few of the solid branches; now for the ornamental. 'Long time ago' she went as far as syntax, read Murray's Geography and through two rules in Pike's Grammar. Could find 6 states on the Atlas, could read, and you see she can write. Can–no *could*–paint roses, butterflies, ships &c, but now she can paint houses, whitewash fences, &c.

Now for her terms.

Her age is none of your business. She is neither handsome nor a fright, yet an *old* man need *not* apply, nor any who have not a little more education than she has, and a great deal more gold, for there must be $20,000 settled on her before she will bind herself to perform all the above.

Not until 1855 and the opening of the railroad across the Isthmus of Panama did the shortage of women end, though no account should omit the mainly Mexican settlement of Sonora in the southern mining district. A Canadian, William Perkins, not only loved the romantic area and the gaiety and colour that Mexican influence brought to the town, but had a high opinion of the Mexican–and French–girls. In this hedonistic paradise a Mexican woman would not marry a heretic, but would live with him as long as she was treated well, or felt like staying, and so would French girls. Surely Mr Perkins was generalising? The contrast in his eyes between one of the first American women to reach Sonora, Mrs Gunn, the 'bonneted, ugly, board-shaped specimen of a descendant of the Puritans' and the 'rosy cheeked, full-formed, sprightly and elegant Spaniard or Frenchwoman' was striking 'even with the full knowledge of virtue' and 'the dangerous fascination of vice'. Perkins was clearly being spoilt in Sonora: in many mining camps Mrs Gunn would have been welcomed, if not with open arms, then with admiring stares.

In Sonora, Sacramento and every other town, the merchants were the class most likely to succeed, given reasonable financial acumen. Some started their rise to fortune as storekeepers of the sort we have already met. Others began in less conventional ways. Domenico Ghirardelli hit the millionaire's trail by selling chocolates and sweets to miners. Few of the Californian millionaires of the second half of the 19th century had ever actually panned for gold during the rush, and merchants and financial wizards like Collis Huntington, Leland Stanford and the de Youngs were dubbed 'grocers' by their rivals, which no doubt failed to spoil their pleasure in their swelling millions. Stanford was perhaps the greatest of them, becoming President of the Central Pacific, superintending its construction, and, later, rising to be Governor of California and a U.S. Senator.

While merchants of soaring ambition worked night and day to make fortunes, ordinary miners, ambitious only for gold, needed entertaining. Entertainment in the very early days often meant (apart from men dancing with each other) gambling, racing, and pitting bulls against bears and humans against bulls. However, there was soon a flourishing theatrical and musical industry in the goldfields. Sacramento had its first theatre in 1849 and San Francisco's first opened the following year. By 1854, the city had several opera companies, to the huge delight of the growing Italian population and many others, for this was an age when opera was popular music. One singer, alas, took to drink after her powers began to fail. She was Madam Biscaccianti, who sunk to the Union Saloon, where 'customers did not mind if she leaned unsteadily against a wall or a table.'

Wandering minstrels and entertainers were all the rage, but most popular of all turns were actresses, any actresses. Actors, unlike the ladies, had to reach a certain standard or else they would get the bird, or vegetables thrown at them. America's greatest actor, Edwin Booth, played in California in the early 50s, aged only 19, and gave the miners his Shylock, Lear, Iago etc., but the phenomenon of the period was the notorious Lola Montez, born in Ireland in 1818, whose conquests had included Ludwig I of Bavaria before she sailed for America in 1851 to astound Californians not so much by her dance, in which a woman fought off spiders, but by her very presence. Her lack of talent was soon apparent, however, and the finest thing she did in the Golden West was to teach a miner's daughter named Lola Crabtree, a pretty girl who rose from child star to become a great comedienne. Child stars were hugely popular in the sentimental gold rush era. 9-year-old Ellen Bateman played Hamlet, also Richard III, her sister Kate, who was later to act with Irving, taking the part of Richmond, while those who liked their ladies larger could enjoy the Blonde Burlesques, led by the English music hall star, Mabel Santley, and other charmers. Comedy–in plays, minstrel shows, and entertainments generally–was more popular than all but the best serious offerings. The miners saw too much tragedy in real life. However, ripe Victorian melodrama often went down well, and sentimental songs of home always did.

Less harmonious was the sound of guns, though in the opinion of Andrew Rolle, California's state historian, the amount of claim jumping has been exaggerated, as has the lawlessness of life in the diggings. Even in the magical first year of the gold rush, before the frustrations of some of those who failed to find gold expressed itself in crime and violence, there were occasional incidents, and in January 1849, Edward Buffum was present at an early display of lynch law.

A Mexican gambler named Lopez had his trunk rifled one night by five armed men, who were soon caught. A citizens' jury ordered 39 lashes apiece and a large crowd enjoyed the spectacle on Sunday morning. Then three of the men, two Frenchmen and a Chilean, were accused of a previous robbery and attempt to murder. They were too

A mixed bunch of miners using a sluice at the head of Auburn Ravine in California

weak with their lashing to attend their trial by a crowd of 200:

> The charges against them were well substantiated, but amounted to nothing more than an attempt at robbery and murder; no overt act being even alleged. They were known to be bad men, however, and a general sentiment seemed to prevail in the crowd that they ought to be got rid of... 'What punishment shall be inflicted?' was asked. A brutal-looking fellow in the crowd, cried out, 'Hang them.' The proposition was seconded, and met with almost universal approbation. I mounted a stump, and in the name of God, humanity and law, protested against such a course of proceeding; but the crowd, by this time excited by frequent and deep potations of liquor from a neighbouring groggery, would listen to nothing contrary to their brutal desires, and even threatened to hang me if I did not immediately desist from any further remarks... I ceased and prepared to witness the horrible tragedy. Thirty minutes only were allowed the unhappy victims to prepare themselves to enter on the scenes of eternity. Three ropes were procured, and attached to the limb of a tree. The prisoners were marched out, placed upon a wagon, and the ropes put around their necks. No time was given them for explanation. They vainly tried to speak, but none of them understanding English ... Vainly they called for an interpreter, for their cries were drowned by the yells of a now infuriated mob. A black handkerchief was bound around the eyes of each; their arms were pinioned, and at a given signal, without prayer or prayer-book, the wagon was drawn from under them, and they were launched into eternity. Their graves were dug ready to receive them, and when life was entirely extinct, they were cut down and buried in their blankets. That was the first execution I ever witnessed. God grant that it may be the last!

Few accounts of a lynching are more memorable than Buffum's, the ex-journalist and soldier turned miner and author, who was also one of the first to glimpse the boundless future of California. Naturally, some of the very worst cases concerned Indians. In 1852 a chief's son was hanged because he had 'insulted a young white woman with an indecent gesture'. Naturally, nothing ever happened to white men who raped Indian girls unless the Indians themselves could take revenge on the right people on rare occasions.

There were no jails in the mining districts except in the few camps which showed evidence of becoming towns, and so punishments were basic: the lash, banishment or death. Sometimes death could be even more basic than a lynch-law hanging, as when a murderer was caught in the act and battered to death with pick handles within 100 yards of his crime. One Mexican girl, who killed an aggressive, foul-mouthed American, was hanged by a mob, though pregnant, and a doctor who spoke up for her was almost strung up as well.

As the gold-fields got more crowded, the easy-going attitude to claims gave way to strict regulations in many places, which helped cut down crime. At Springfield in 1852, the following articles were included in a poster proclaiming the local laws and boundaries:

ART. 1 A claim for mining purposes within this district shall not exceed one hundred feet square to each man.
ART. 2 That no man within the bounds of this district shall hold more than one claim.
ART. 3 That each and every man holding a claim within the bounds of this district shall work one day out of three, or employ a substitute; otherwise such claims shall be forfeited.
ART. 4 That any persons holding claims with dirt thrown up or excavations made on such claims, such claims shall

The last of the Cariboo camels, a form of transportation in the British Columbian gold rush which died out because packers objected to the animals' biting, kicking and smell

be governed by articles 2 and 3.
ART. 5 That each and every man holding a claim within the bounds of this district, shall designate such claims by erecting good and substantial stakes at each corner of their claims, or dig a ditch around said claim, with a notice signed by each person or individual of a company holding them.

At least the would-be claim-jumper knew exactly what he was letting himself in for, especially as there was a mention of a posse in Article 9. There was a vague unwritten law on ordinary, as opposed to land, robbery, the custom being to string a man up if he stole $100 worth of anything, but instant executions occurred for far more trivial sums.

If the amount of lawlessness in the gold-fields has been exaggerated, there can be no doubt that booming San Francisco, whose population shot from 800 in March 1848 to 25,000 in 1850, and topped the 50,000 mark in 1855, was as wild as it was booming, and expensive. As well as the 'Yankees of every possible variety' that Bayard Taylor noted in 1849, along with 'native Californians . . . Chileans, Sonorians, Kanakas from Hawaii, Chinese with long tails, Malays' and others, there were two notorious gangs. The 'Hounds', otherwise known as 'Regulators', were mostly youthful ex-soldiers disbanded in California, while the 'Sydney Ducks' were ex-convicts of Queen Victoria, who had arrived from Australia, and whose behaviour soon had the respectable residents saying: 'The Sydney Ducks are crackling.'

The Regulators' worst excess was against the local Chileans. On July 15, 1849, after a so-called patriotic parade and a drunken tour of saloons, they descended on the tent-town known as Little Chile, beating up and shooting the inhabitants. Miraculously, there were no actual deaths.

This atrocity roused respectable folk to action, notably our old friend Sam Brannan, who since his 'Gold from the American River' days had become an amazingly successful property speculator, netting $160,000 dollars in shrewd business deals in 1849. He and his friends organised a 230-strong committee, formed a citizens' court, tried the leading Begulators, and banished them from San Francisco.

For a while crime was kept to reasonable proportions by boom-town standards, though those who exploited their fellows economically and politically cannot be included as they were never regarded as criminals. The city concentrated on having a good as well as a prosperous time. Some miners felt it their duty to leave San Francisco totally broke after a stay in the city, and there were plenty to help them spend their money. The price of gold there sometimes dropped to only eight dollars an ounce, though when the U.S. branch mint opened in 1854, the sensible miners went there knowing that they would get the official 16 dollars.

On arrival, a bath and a good meal were on the agenda, possibly a new suit of clothes, and the meals, especially in the foreign restaurants, were fabulous for men used to pork and beans, and often fabulous by any standards. Gambling and drinking in halls that grew more splendid every year were prime attractions, especially as female company increased by the month. Foreign girls were especially popular, and charged fortunes for their favours, which they could never have got at home: an ounce of gold to sit and

Lola Montez, the Irish-born adventuress who managed to take in the Australian as well as the Californian gold rushes

drink with a miner, 30 ounces or so–some 400 dollars–to sleep with him. The most successful bordellos were promptly rebuilt more splendidly than ever after the fires that regularly swept the mainly wooden, jerry-built city–there were six fires between Christmas 1849 and June, 1851–and only the most virtuous women found it easy to avoid being swept up in an atmosphere which at the least was exhilaratingly amoral.

Every sort of theatrical entertainment flourished in San Francisco, where, in the 1850s, some 907 plays were staged, also 84 extravaganzas, pantomimes and ballets, 48 operas and 66 minstrel shows. There were also race tracks and amusement parks, and there were duels, some formal, some drunken, and plenty of other events to keep the readers of the *California Police Gazette*, with its lurid woodcuts, happy. At Madame Reiter's Bagnio, for instance, a patron named Frank Lines had been relieved of his wallet, so returned later 'to clean out' the establishment by shooting the hostess. At lesser establishments on the notorious Barbary Coast, 'pretty waiter girls'–it was not done to use the word prostitute–were adept in parting customers from their money first on drink, then by selling their favours at exorbitant rates, finally taking what was left by picking their pockets. There were few to return, like Frank Lines, to mete out instant justice, and the fact that one notorious house was nicknamed Murderer's Corner speaks for itself.

Before rejoining the Sydney Ducks, a few gems from the Police Court Column of the *Daily Alta California* of 1852 might be in order, not only to indicate minor types of crime, but because of the casual anti-foreign attitudes expressed in Anglo-American reporting in gold rush days. 'John Gomez, a red-shirted, squalid looking greaser' is sent down for ten days and fined $25 for beating up his wife, and a 'greaserita named Carolina and an American named Hyde' are fined $20 each and 'put in the jug for ten days' for 'fighting desperately on Dupont Street'. John Briggs fared better, being found only 'comfortably drunk on Long Wharf' and was discharged on promise to reform, while the insensible George Ditz was 'taken up in a hand-cart and emptied in the police station' where he was fined $5.

A year before these minor events the first great vigilance committee had come and gone. In 1851, law enforcement and law courts were hopelessly inadequate. 'How many murders have been committed in this city within a year?' asked the by-no-means inflammatory *Alta California*. 'And who has been hung or punished for the crime? Nobody!' It went so far as to suggest that the threat of lynch law might be the only remedy. And it was not simply weakness that prevented adequate law and order, but graft and corruption in the police and the courts alike.

The year started livening up in February, when a leading citizen named Jansen was beaten up and robbed in his store. The alleged culprits were caught and tried, when Jansen identified them, but during the trial a mob of several thousands broke in and almost hanged the men before agreeing to appoint a special court, which failed to agree. Almost hanged once more, the men were tried again and convicted, one later escaping, the other being freed when the true culprits were discovered after one of them was captured for another crime.

This classic example of the way innocent men could be wrongly accused was swiftly forgotten when the Sydney Ducks–no innocents they–went too far. Though only one of the gangs of the city, these slum-based menaces, ex-convicts, ticket-of-leave men, and, apparently, some of them never convicts, were rarities in gold rush California in that they had simply come to plunder by robbery and arson, the latter being a splendid opportunity for looting. On the night of May 3, just a year since a major fire, one of the Ducks made good the boast of the gang's leaders over the past few weeks to set the city on fire.

Shortage of water made the fire a bad one, during which the Ducks began to loot. Several were killed by furious citizens, but 2,000 buildings were destroyed, and a voluntary patrol to help the police during fires was formed. Then, on June 10, 'The Committee of Vigilance of San Francisco' was formed, with some 200 members pledged 'to watch, pursue, and bring to justice the outlaws infesting the city, through the regularly constituted courts, if possible, through more summary course, if necessary ... no thief, burglar, incendiary, or assassin, shall escape punishment, either by the quibbles of the law, the insecurity of prisons, the carlessness or corruption of the police, or a laxity of those who pretend to administer justice.'

A former Australian convict named John Jenkins chose this moment to steal a strongbox from a shipping office, and two peals of the city's fire bell rang out to assemble the Committee. When he was caught, he was handed over to the vigilantes, judged by Sam Brannan, and early the following morning, after a rescue attempt by his friends had failed, was hanged. 183 men signed a public statement claiming that they were all 'equally responsible for the first act of justice that has been dealt to a criminal in San Francisco since California became a State of our Union.'

On July 11, 'English Jim' Stuart was arrested, confessed to various crimes, for one of which the vigilantes had planned to hang an innocent man, and was hanged. In a three-month period of existence, the vigilantes in all hanged four men, whipped another, deported fourteen, banished one, discharged 41 and handed fifteen over to the authorities. There were 707 men on the strength when the committee stood itself down, though a mere two dozen

Still Another.

BANK OF SANTYCLAUS,
Upper California. May 17, 1849.

EDDYTURS OF THE SUNDAY TIMES,—I bleve I told you in my last letter 'bout Gineral Persevere Smith's proclamashuns agin furrenirs.—There's bin a diffikulty between him an' major part of the minors sense then. He wanted to rivet a tacks on the furren minors, and they woodent submit to the imposition. O, there was an orful time! He told 'em he was the Konkeror of Contraries, and woodent submit to no contradiction. He also shode 'em the authority of the government. an' ast 'em if they'd dispute that? They said no, they woodent dispute nuthin; but they'd dig as much gold as they'd a mind too, and if he interfered with their siftin, his own sand 'ud soon be run.

The rush to the diggins continues, and the provisions is getting skarse, so much so that fat men begins to be regarded with avarishus eyes. Infants, if plump, woodent be safe, an some of the fellers look at the two or three young wimmen we have in the settlement jest as if they wanted to eat 'em. Appetite's no respecter of persons, as a young man from St. Joseph's who ett his grandmother on the plains, remarked to me the other day. "The old ooman," says he, "was dry. very dry, but there's no sarse like hunger, and without intendin to make game of old age, I must say, she tasted like venson."

The injuns has been troublesome lately. The licker havin given out, the darned red skins refused to dig, and we had to lam 'em; whereupon they got sausy and fit with us. In coarse we used 'em up, and now the cussed ungrateful devils wont come near us. Aint it too bad, seein how weve done for 'em. But it's so all the world over. I've a good mind never to make an aboridgines drunk again, or act the part of a chrystin by any o' the vile mule stealing, lasso throwing, skull skinnin, copper heads again while I live.

We have noose by the Californy steamer that a mense quantity of grub and spirits is cumin out from the States. If we get it in time we mean to have a glorious blow out on the Fourth of July. Although we have found a temporary hum in Californy, we feel cheap when we reflect upon the dear ones we left behind. Sometimes a man of family dreams that he is agin with his wife and children, but in the morning the delushun vanishes, and he finds himself still in a Pacific State. Others who left the partners of their buszums, fancyin they had cause for jealousey, would willingly dubble the Horn to clasp them in their arms wunce more.

As I menshund in my last letter I have akwired suffishent welth, an long for the sweets of connubial bliss. Love is stronger than Cupidity. At present, my effeckshuns embrace wimmen generally, when I get hum they will soon come to a focus.

Please bear my wife in mind—I mean the one you are to select for me. I will write and let you know when I expect to be at Panyma, and would like her to jine me there, and be united at wunst without waiting till we get to the United States. Yours,

A DISBANDED VOLUNTEER

☞ There is but one step from the sublime to the ridiculous as has frequently been acknowledged. We can imagine the feelings of a bellicose individual whose ire has been so excited as to cause him to challenge another man to mortal combat upon receiving a reply as ridiculous as the following, which emanated from one Carl Ambruster, a worthy sausage maker of St. Louis.

I, Carl Ambruster, sausage maker in chief to

[Fro
Import
what it is a
have been a
tion. We
features.
In Februa
citizens of tl
east of the S
vention wou
City, on the
ing into consi
a territorial o
Accordingly
vention met,
inhabitants of
lying east of
Daniel Spenc
Clayton secre
retary, and I
After seve
was appoint
which they
Congress of
provide.
Commit
Phelps, J
Birnhisel
Charles
The c
the 8th
the co
Thi
pres
the
"
Being
feelin
of th
and
the
boun
nort
lon
sou
ice
Gi
or
th
w
t
dividing ra
waters flo
north from
sin on the
chain of mo
by the divi
the waters
the waters
the place of
by Charles
Senate of
The pow
into three
and Judic
The art
partment
stitution
quired t
ted Stat
constitu
sist of se
ty-five n
In the
for the el
nor, Secre
counts ar
vested in

GREAT LAND ENTERPRISE AT LUNAVILLE!
ROUSING OPPORTUNITY!
Five Hundred Acres on the Sunny Side of the Moon to each Subscriber.
WITH LOTS OF ROCK FOR BUILDING PURPOSES!
When one half the stock is taken, an Atmospheric Engine will be erected in the crater of Popocatapetl to furnish refined air to the settlers, and a Steam Squirt will be placed on Goat Island to play water on the Moon, so that the inhabitants will have always enough—never too much, and never too little; thus avoiding the drouths and drenchings to which the earth's people are liable.
Balloons also will be provided to start daily from different available points on the earth.
NOW IS THE TIME TO SUBSCRIBE!

A miner using his rocker in Alder Gulch, Montana, in the 1870s

had master-minded the operation, which had broken the power of the Ducks and given a grim warning to other criminals that was not lost on most of them.

Most Western newspapers backed them, though eyebrows were raised in some quarters in the East, and modern historians have tended to look sceptically at the old herioc-noble image of the propertied vigilante. It is only fair to point out that, for all the inherent dangers of their actions, the vigilantes were coping with an almost intolerable situation.

We have seen how their actions were copied in various ways beyond San Francisco. In 1856, they reactivated themselves in the city, with plenty of justification, for the memory of the hangings began to fade, and the murder rate reached a new peak. So did official corruption. The latter is harder to estimate, but some 1,000 unpunished murders between 1849 and 1856 give some indication of just how wild San Francisco was.

Two murders in particular brought about the new vigilance committee. The first happened in November, 1855, when General Richardson, a federal marshal, was murdered by a gambler, Charles Cora, whose mistress had quarrelled with Richardson's wife at the theatre. A gross mis-trial occurred and the gambler was freed, then, in May, 1856, a crusading editor named James King was shot down by one of his targets, a powerful politician named Casey, who had done time in Sing Sing, and who was just the sort of corrupt local official that the ex-vigilantes and other reasonably reputable citizens objected to. A new committee was rapidly formed, nearly 10,000 strong within a fortnight, and with an executive of 39. Discipline was strict and nothing the city officials could do–they tried calling out the militia, the Army and the Navy–could stop them, especially as their leader, William Coleman, told the Governor of California their plans and he simply said: 'Go to it, old boy!'

Cora had already been arrested for a retrial just before editor King was murdered. Indeed King, assuming the gambler would get off again, had told the city's gamblers and whores to 'rejoice with exceeding gladness.' The vigilantes grabbed Cora and Casey, tried them on the day that the mortally-wounded King finally expired, and on the day that some 20,000 watched his funeral, had the two murderers hanged in front of another vast crowd as the bells still tolled for the editor who had fought his last crusade.

After banishing a number of other undesirables and hanging two more, this vigilance committee, too, abolished itself. High-handed these vigilantes certainly were, but they nevertheless achieved their main ends, for San Francisco was to enjoy some years of reasonably good government entirely because of them. And their willingness to stand down hardly suggests a gang of bully boys.

California's were not the richest strikes even in the United States in terms of money made, but its gold rush was the greatest of them all. The Argonauts not only set the Golden State on its rocket-like rise to a prosperity which is still growing today, they triggered off America's rise to a world, then *the* world, power. Westward emigration had been crucial but small scale in the early 1840s, now it spiralled, and the transcontinental railroad set the seal on the process, along with the countless thousands of immigrants. Gold fever spelt national prosperity.

The first seven years of the rush yielded more than $350 millions worth of gold, the *annus mirabilis* being 1852, when $81,294,700 came out of the mines. The next year some 100,000 men were active and $67 millions were made. After that, the fall-off began, and the individual gave way gradually to the mechanised machines of the mining companies.

Some quarter of a million people from all over the world had come to California. That some were to become fantastically rich and most were doomed to be unlucky cannot disguise the fact that many were not prepared to work hard enough, for in '48 and '49 especially, a hard worker had to be very unlucky to gain next to nothing. Later finds, as we shall see, were wildly over-exaggerated, often by brazen liars, but only rarely California's, where the problem was accurately summed up by a physician named McCollum: 'The abundance of gold in California has not been so much overrated as the labour of procuring it has been underrated.' And the admirable reporter Bayard Taylor noted that those 'who, retaining their health, return home disappointed say that they have been humbugged about the gold, when in fact they have humbugged themselves about the *work*.' Edward Buffum flatly stated that for the man with a strong constitution, hands used to work and heart not broken by disappointments, 'there was never a better opportunity in the world to make a fortune.' That all those quotations come from the early bonanza years does not lessen their accuracy. Some learnt quickly, others not at all.

The most admired men were miners who made and kept fortunes, not the 'grocers'. Daniel and John Murphy made $1.5 millions in 1848 and later respectively owned three million acres of cattle land and entered politics, while, amongst others, at least one friend of the unfortunate Sutter, John Bidwell, struck it hugely rich, became a major landowner and, finally, a Presidential candidate.

The average gold-seekers either headed for other fields in the States, British Columbia or Australia, settled down in California, or went home, having experienced the great adventure of their lives. In retrospect, even the least fortunate began to reminisce nostalgically about the good old days, or so it would seem. Had they not seen the elephant?

Miners at work in Deadwood, South Dakota, in 1876

The Diggers

I felt myself surrounded by gold. . . .
Edward Hammond Hargraves

One of the visitors to Sutter's Mill in the summer of 1849 was an English ex-sailor, Edward Hargraves, who called to buy lumber from the famous, if ill-fated, James Marshall. As the two men leant against some wood, Hargraves mentioned he was from Australia. Born in 1816, he had sailed there as a boy of 16.

'Then why don't you go and dig among your own mountains?' said Marshall, 'for what I have heard of that country, I have no doubt whatever that you would find plenty of it there.'

Hargraves studied the landscape as he mined–not particularly successfully–and became convinced that it bore a striking resemblance to land he dimly remembered in New South Wales. He decided to go home to his family and to see if he remembered correctly.

'There's no gold in the country you're going to,' said an American as he left, 'and if there is, that darned queen of yours won't let you dig it.'

Allegedly, Hargraves removed his hat, took up a theatrical stance, and proclaimed: 'There's as much gold in the country I'm going to as there is in California, and Her Most Gracious Majesty the Queen, God bless her, will appoint me one of her Gold Commissioners.' She did.

However, it was 'with an anxious heart' that he landed at Sydney in January, 1851, especially as everyone on the voyage had thought him mad. He set off alone on horseback to cross the Blue Mountains, stayed a night of the journey at Bathurst, and finally reached his destination, Guyong, which he remembered so hopefully from eighteen years before.

As far back as 1823 it had been known that there was gold in Australia, though little notice was taken of the finds. That year a convict picked up a lump near Bathurst, but was flogged by his officer, who decided he had been melting stolen goods. And the same year Assistant-Surveyor James M'Brian recorded in his field book that he has found 'numerous particles of gold in the sand and in the hills' fifteen miles east of Bathurst.

Other finds occurred, most notably one by a Polish scientist and explorer, Paul Edmund Strzelecki, in 1839 and the Rev William Clarke several years later. The Pole was implored by the Governor of New South Wales, Sir George Gibbs, to keep his discovery to himself in case the Colony became unsettled and the convicts rebelled, and Clarke was urged to do the same to prevent blood-letting. Both men compromised, Strzelecki mentioning gold when his journal was published, but stating that there was not enough to repay extraction–he thought the very opposite–and Clarke shipping his specimens to Sir Roderick

Edward Hammond Hargraves discovers gold in New South Wales, watched by young Lister

Murchison, the great geologist and explorer. In 1844, Murchison addressed the Royal Geographical Society and compared the rocks of the Blue Mountains to those of the Urals, where gold had been found, and announced: 'There was every probability that the one would be found to be as rich as the other was known to be in precious metals.'

That not the slightest sign of a gold rush occurred is strange, though the fact that by British law any gold or silver found was Crown property was hardly encouraging. Transportation of convicts to New South Wales had ceased in 1840 (they were sent to Tasmania and Norfolk Island until the '50s), though the authorities can hardly be blamed for fearing an uprising. Criminals had been sent to the colony on conditional pardons and as ticket-of-leave men to the growing fury of most of the population, rightly resenting the Government in London regarding what was now a homeland as a dumping ground for undesirables. And with the convicts still there and bushrangers flourishing, Sir George's attitude and fear for his throat was understandable.

Why then did Hargraves, whom we have left spending the night at Guyong, encounter little opposition? Partly because there was something of a slump in the colony, but, especially, because it had lost so many men to California. Fortunately for Australia, Hargraves wanted to come home. His find, and the subsequent events, were to raise Australia's population from 437,665 in December, 1851 to 1,168,149 in December, 1861. Victoria, where the biggest finds were to be made in that decade, shot from 97,489 people to 539,764.

The Australia to which Hargraves had returned contained sixteen times more sheep than people. Sheep-farmers, branded as squatters by successive governments in London which claimed that all land under British rule was Crown property, had finally made their point to the Colonial Office, and were now the most influential section of the community, along with a colonial administration that was mainly authoritarian. There were plenty of tensions between the rising middle classes in the few cities and the ascendency of landowners and government, not least because sheep-owners were eager to have transportation brought back, with the resulting cheap labour. Essentially, Australia was a backwater, for there was little feeling for Empire in the Britain of 1851, and the emotional hey-day of Imperialism was a generation away. Some British politicians behaved as if an American Revolution had never happened. Then Hargraves triggered off one of the three key events in Australian history, the other two being the World Wars.

He set out from the inn where he had stayed the night, and where he had been told about a local gold find, and with him went the son of the house, young Lister, aged 18. It was February 12, 1851.

A few hours later, they were at Summer Hill Creek, a tributary of the Macquarie River and soon they had reached his land of heart's desire, the spot that looked so like the country round Sacramento. He could not be mistaken and felt himself surrounded by gold:

> My guide went for water to drink, and, after making a hasty repast, I told him that we were now in the gold-fields, and that the gold was under his feet as he went to fetch the water for our dinner. He stared with incredulous amazement, and, on my telling him that I would now find some gold, watched my movements with the most intense interest. My own excitement, probably, was far more intense than his. I took the pick and scratched the gravel of a schistose dyke, which ran across the creek at right angles with its side; and, with the trowel, I dug a panful of earth, which I washed in the water-hole. The first trial produced a little piece of gold. 'Here it is!' I exclaimed; and I then washed five panfuls in succession, obtaining gold from all but one....

Hargraves was wildly excited, but still able to make a suitable speech to his audience of one. It was later to be mocked as Stanley was mocked for his 'Dr Livingstone, I presume?', but was undeniably worthy of the occasion–'This is a memorable day in the history of New South Wales. I shall be a baronet, you will be knighted, and my old horse will be stuffed, put in a glass-case, and sent to the British Museum!' Without reaching such giddy heights, Hargraves did well enough, in striking contrast to Marshall. Apart from becoming a Crown Commissioner of Crown Land, he was voted £10,000 by the New South Wales legislature, worth perhaps a quarter of a million in today's terms.

Meanwhile, he panned in other areas of Ophir, as he christened his wonderland, taught his companion and another youth how to make a rocker, then returned in triumph to Sydney. The Colonial Secretary Deas Thomson, and his colleagues were doubtful at first, suspecting that Hargraves had 'planted' Californian gold, but the official geologist confirmed the truth. In April, Thomson had said to Hargraves: 'If this is a gold country ... it will stop the Home Government from sending us any more convicts, and prevent emigration to California; but it comes on us like a clap of thunder, and we are scarcely prepared to credit it.' Now he was convinced.

Hargraves was lucky, of course. Although not the first to find gold in Australia he yet ranks as the official discoverer. But if he was lucky in his timing, he deserved his good fortune because he had a plan–a vision–and he had the inspired dedication to carry it through.

By May 19, four days after Deas Thomson had announced the discovery publicly in the *Sydney Morning Herald*, there were 400 people on Summer Hill Creek, and Bathurst was almost emptied of people, except for smiths doing a roaring trade making pick-axes. Every day the numbers at the diggings grew. The pleasant little gulley where Hargraves had made his first find was already almost ripped to pieces, and the creek nearly blocked by gravel heaps.

On May 28 and 30, Colonel Mundy, Deputy Adjutant General of the Military Establishment, drove to the races at Homebush, ten miles out of Sydney:

> I counted nearly sixty drays and carts, heavily laden, proceeding westward with tents, rockers, flour, tea, sugar, mining tools, &c., each accompanied by from four to eight men, half of whom bore fire-arms. Some looked eager and impatient–some half ashamed of their errand–others sad and thoughtful–all resolved. Many, I thought, would never return. They must have thrown all they possessed into the adventure; for most of their equipments were quite new–good stout horses, harness fresh out of the saddler's hands, gay-coloured woollen shirts, and comforters, and Californian sombreros of every hue and shape.... The mind could hardly reconcile a thoroughly English highroad, with toll-bars and public houses–thoroughly English figures travelling on it to a country race-course.... with the concurrent stream of oddly-loaded drays and other slow-moving vehicles, piled with business-like stores and unfamiliar utensils, and escorted by parties of no less English men, armed to the teeth, clad in a newly adopted dress, utterly indifferent to and apart from the merry scene of the race course, and carrying with them a dogged, resolute and abstracted air...

The whole of New South Wales was soon stricken with gold fever and, as in California, ships were promptly deserted, store-keepers closed, or were forced to close, their shops, businessmen and their clerks left their offices, and all headed for the gold-fields. A local touch was that shepherds hastily deserted their flocks.

Australia got its first taste of instant democracy. If the Californian gold-fields had seen a great levelling of the classes, how much more inevitably occurred in a Crown colony of the 1850s. The Governor himself is alleged to have had to groom his own horses, while a ship's captain, feverishly trying to find a crew for his next voyage, came upon a drunk and asked him if he would be so good as to join his crew. Instead, the apparent down-and-out pulled out a wad of money and said: 'I'll buy your barque, cap'n, and ship you with her!'

Ophir contained everyone from magistrates, doctors and lawyers to cobblers and ex-convicts, while one tailor, whose entire staff of ten walked out on him, headed for the diggings himself and became their cook. There was even a 'real live lord on his travels.'

By the end of July, there were some 3,000 on the gold-fields, including many from other colonies, and Colonel Mundy went up to inspect the scene:

> As we topped a ridge, the last of a series I thought interminable, my companion suddenly said, 'Stop and listen'. I pulled up my horse and heard as I imagined the rushing of some mighty cataract. 'It is the cradles', said he: and so it was–the grating of the gravel or rubble on the metal sifters of five hundred rockers.... There was no pause nor the slightest variation in the cadence as it floated up to us on the still air.

This was at the diggings on the Turon River, where there had been another big find. The miners had to pay thirty shillings a month licence money for the privilege of digging to the Commissioners of Crown Lands, and in New South Wales at least, the system worked quite well with a humane commissioner who allowed miners a week's credit. £3 4s. an ounce was the going rate for gold, and the average earnings at this stage were from 10s. to £1 per man per day. The many who did not make this amount, enough to live on in a modest way, had a chance of being employed by those who *were* making money.

Some were making plenty. The lucky few found nuggets clustering to the roots of shrubs or simply lying on the ground, some got it the hard way, and a handful had sensational finds. The most sensational of all, the Kerr nugget, was found in July when mid-winter weather was discouraging all but the most hardy. 'Bathurst is made again! The delirium of golden fever has returned with increased intensity. Men meet together, stare stupidly at each other, talk incoherent nonsense, and wonder what will happen next,' exulted the *Bathurst Free Press* of July 16.

Three aboriginal youths employed by a Doctor Kerr had found a huge mass of gold encased in quartz, and he broke it into several pieces to fit into his saddlebags. Its arrival in Bathurst caused a sensation, and the doctor made £4,000 out of it, the market value for 106 pounds worth of pure gold, but, as our friend, Colonel Mundy, observed:

> It was, indeed, impossible to avoid lamenting that this unique specimen of virgin gold–rock and ore–had not been removed in a state of perfect integrity from its native bed to Sydney, and from thence to London....
> Looking at the monster lump in a speculative light, Mr

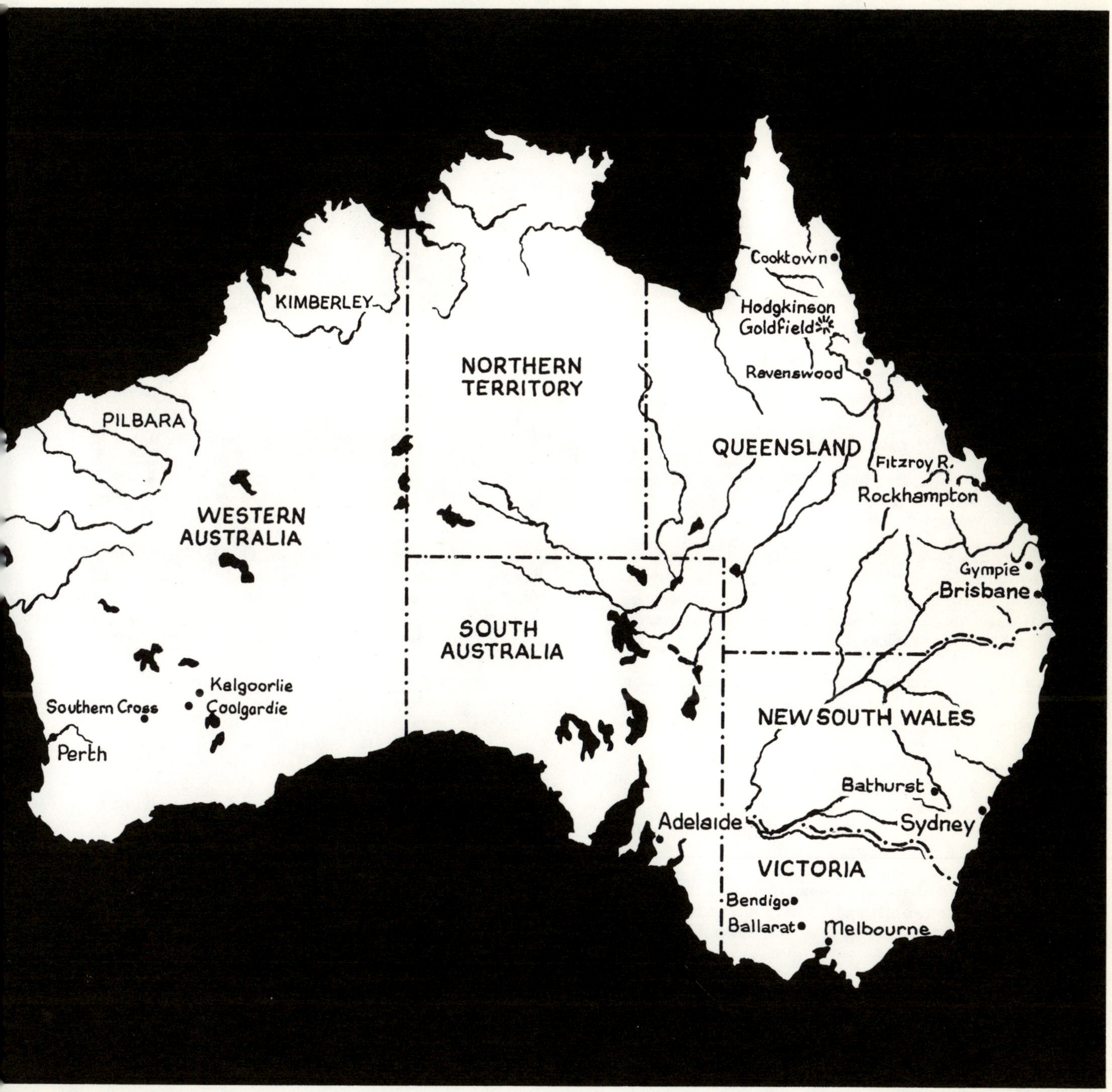

The Australian gold rushes

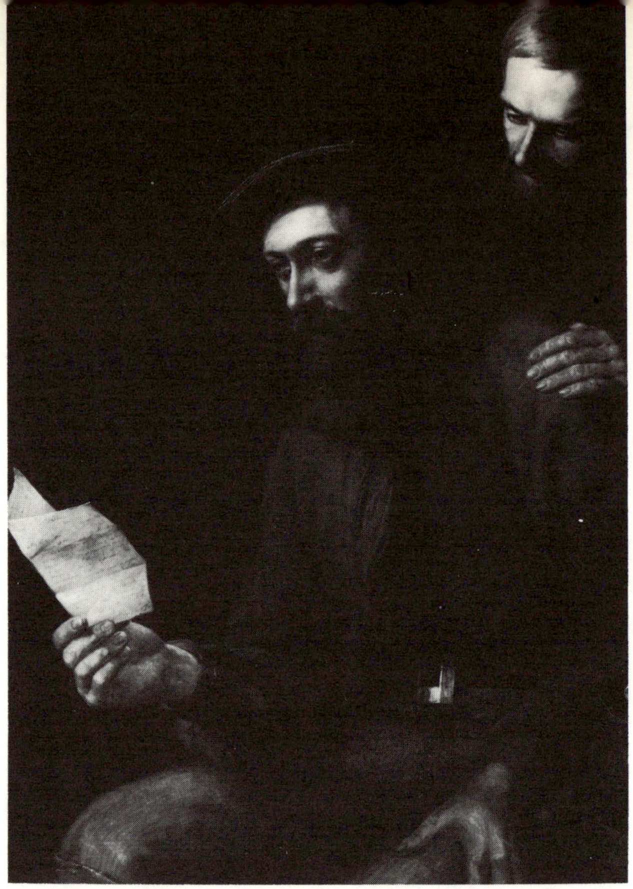

Barnum would have realised 50,000 to 1 in a couple of years by exhibiting it round Europe and America with the black fellow who found it, and the saddle bags in which it was abducted, and would have sold it afterwards for at least twice as much as Dr Kerr got for it.

At least the doctor treated his black fellows well at a time when they were at best despised. He gave them two flocks of sheep, a dray, and a team of bullocks, and other presents.

While New South Wales was enjoying its gold rush, comparatively free of tension between the Crown and the diggers, and remarkably free of crime, the colony of Victoria, recently separated from the mother colony of New South Wales, was being depopulated of its small numbers of settlers, then some 97,000, with 23,000 in Melbourne. A 'Gold Discovery Committee' was therefore rapidly set up by leading citizens, offering a reward of £200 for the first person to strike gold within 200 miles of the city.

In fact, several major finds occurred from July onwards near Ballarat and Bendigo, and by November some 25,000 diggers were hard at work in Victoria. By the end of the year nearly £9 million had been extracted. Clearly, the Victorian fields were on a far larger scale than those in New South Wales.

The prize pickings were sensational. One Ballarat digger dug five feet and found 'gold was so thickly sprinkled that it looked like a jeweller's shop.' Another made £1,800 in a day. The Welcome Stranger nugget weighed 2,280 ounces. In ten years Victorian fields were to produce £100 millions.

The news of the Australian finds reached the outside world by the end of 1851, and every corner of the world sent its quota of adventurers, especially Britain, America and China. 100,000 new arrivals swelled the population in 1852, and by 1855, Victoria boasted more people than had been in all Australia before Hargraves found himself surrounded by gold.

Much of what happened in Victoria had previously occurred in California, not simply the depopulation of cities–Melbourne was almost paralysed for months–but in the hostility shown to the Chinese. Diggers tossed dynamite into the crackers when the 'Celestials' were celebrating the New Year, and Anthony Trollope later made a typically Anglo-American observation about the Chinese on being shown their quarters at Ballarat: 'A more degraded life is hardly possible to imagine. Gambling, opium smoking, and horrid dissipation seem to prevail among them constantly. They had no women of their own, and the lowest creatures of the street congregate with them in the hovels ... Boys and girls are enticed among them ...' That he was writing about the early 1870s does

Would-be diggers going on board at Melbourne in 1893, Coolgardie bound

Gold diggers in Ballarat, New South Wales, receiving a letter from home. The painting is attributed to William Strutt

not lessen the point.

Looking not unlike the Forty-Niners, except that Australians tended to wear the cabbage tree hat, the diggers sweated and some of them died of disease, of falling into water-filled holes and drowning, but rarely of violence. They also received a visit from the notorious Lola Montez, who provoked even more mixed reactions than she had in California. And, without being exactly well-supplied with women, they undoubtedly were not as deprived as the Forty-Niners. There were many wives and children at the gold-fields, digging and washing for alluvial gold with their husbands, and there were other women, including one seen by Lord Robert Cecil on his travels many years before he became, as Lord Salisbury, Prime Minister of Britain:

> ... we saw a digger in his jumper and working dress walking arm in arm with a woman dressed in the most exaggerated finery, with a parasol of blue damask silk that would have seemed gorgeous in Hyde Park. She was a lady (so the driver told us) of Adelaide notoriety, known as Lavinia, who had been graciously condescending enough to be the better half of this unhappy digger for a few days, in order to rob him of his earnings. These women are no rarities at the diggings.

A gun was fired from the Commissioner's tent in most diggings when it was time to stop work and enjoy a quiet evening in tent or shack, or gamble, or dance–often, as in California, with other men. As for Sundays, a respectable woman named Mrs Clacy, who kept 'house' for her brother and some of his friends in the gold-fields, later wrote:

> Sunday is kept at the diggings in a very orderly manner; and among the actual diggers themselves, the day of rest is taken in a *verbatim* sense. It is not unusual to have an established clergyman holding forth near the Commissioner's tent, and almost within hearing will be a tub orator expounding the origin of evil, whilst a 'mill' (a fight with fisticuffs) or a dog fight fills up the background.

But there was another, darker side to the picture, and not merely the chronic ill-health of so many miners: cramps, colds, rheumatism, bad eyes, diarrhoea and dysentery were the prevalent complaints according to James Bonwick, a digger turned prolific author. It was not even the rise of crime which, as we shall see, afflicted the Victorian diggings from 1852 onwards. It was the diggers' legitimate grievances over the licence fees, and the type of men sent to the diggings to collect them, which was to culminate in armed rebellion.

Puddling for gold at Kalgoorlie in the 1890s. This involved stirring the pay dirt for some time because of shortage of water. The slime was drawn off regularly through a plugged hole

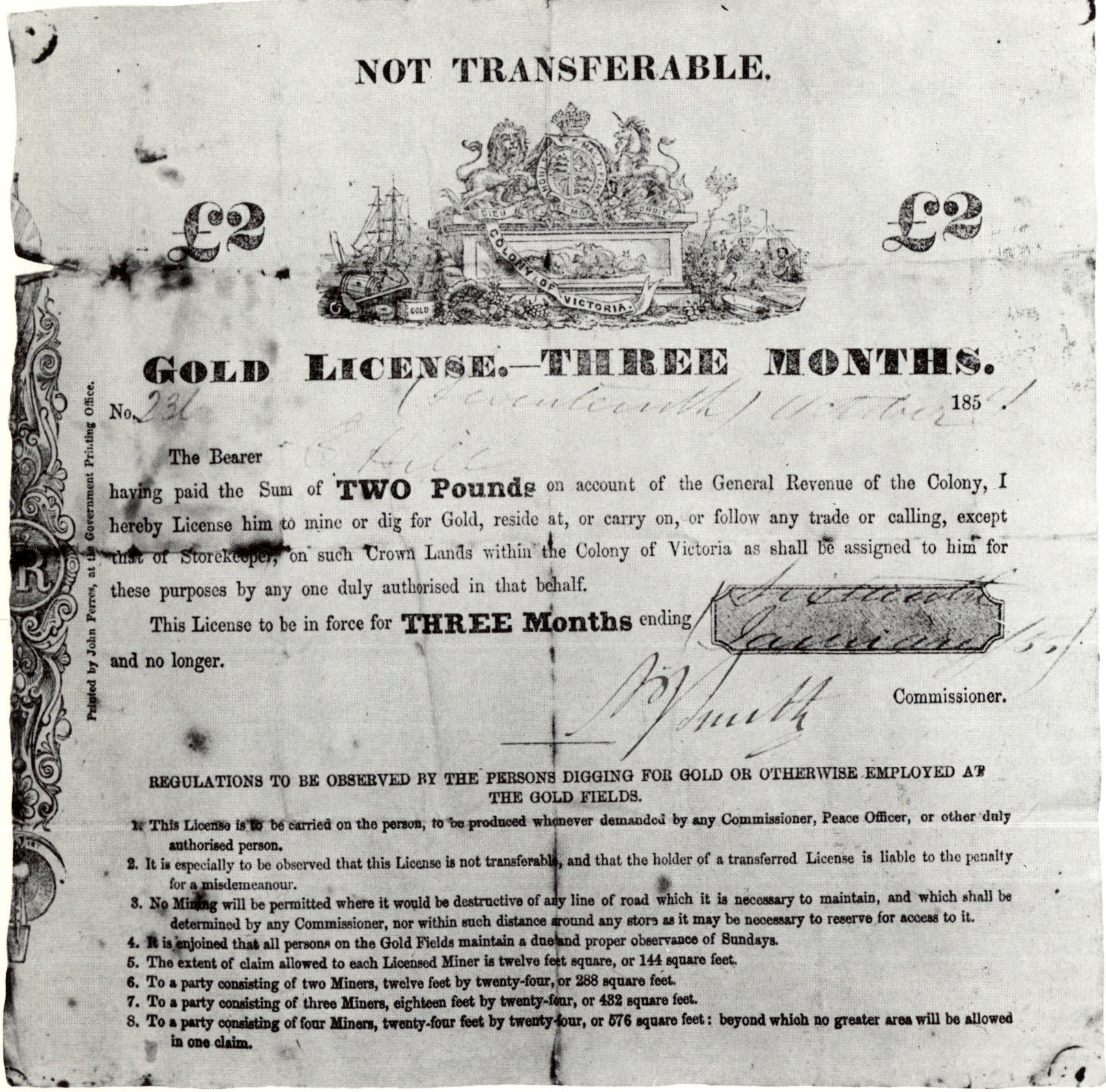

NOT TRANSFERABLE.

£2 £2

GOLD LICENSE.—THREE MONTHS.

No. 231 (Twentieth) October 185

The Bearer

having paid the Sum of **TWO Pounds** on account of the General Revenue of the Colony, I hereby License him to mine or dig for Gold, reside at, or carry on, or follow any trade or calling, except that of Storekeeper, on such Crown Lands within the Colony of Victoria as shall be assigned to him for these purposes by any one duly authorised in that behalf.

This License to be in force for **THREE Months** ending

and no longer.

Commissioner.

Printed by John Ferres, at the Government Printing Office.

REGULATIONS TO BE OBSERVED BY THE PERSONS DIGGING FOR GOLD OR OTHERWISE EMPLOYED AT THE GOLD FIELDS.

1. This License is to be carried on the person, to be produced whenever demanded by any Commissioner, Peace Officer, or other duly authorised person.
2. It is especially to be observed that this License is not transferable, and that the holder of a transferred License is liable to the penalty for a misdemeanour.
3. No Mining will be permitted where it would be destructive of any line of road which it is necessary to maintain, and which shall be determined by any Commissioner, nor within such distance around any store as it may be necessary to reserve for access to it.
4. It is enjoined that all persons on the Gold Fields maintain a due and proper observance of Sundays.
5. The extent of claim allowed to each Licensed Miner is twelve feet square, or 144 square feet.
6. To a party consisting of two Miners, twelve feet by twenty-four, or 288 square feet.
7. To a party consisting of three Miners, eighteen feet by twenty-four, or 432 square feet.
8. To a party consisting of four Miners, twenty-four feet by twenty-four, or 576 square feet: beyond which no greater area will be allowed in one claim.

One of the hated 'licence to dig for gold', which helped trigger off the events leading to Eureka Stockade

A mining family stands outside its shack

Eureka Stockade

We swear by the Southern Cross to stand truly by each other and fight to defend our rights and liberties.
The Diggers' Oath, 1854

'Men are robbed almost every night; tents are cut open; and on the road to town some very horrible outrages have been committed,' wrote a miner in 1852 about conditions around the Mount Alexander diggings. The 'Australian Journalist' who later published this and other letters from unidentified correspondents in his *The Emigrant in Australia or Gleanings from the Gold-fields* (1852?) waxed, indignant on the subject of the breakdown of law and order in one area at least, and, indeed, the Victorian diggings were plagued by crime in a way that the ones in New South Wales never were. It was on a small scale compared with California, but it particularly infuriated honest diggers. As the unknown 'Journalist' wrote:

> All the letters from Mount Alexander dwell upon the lawless condition of the place, and the deeds of rapine and bloodshed that disgrace it. The well-disposed miners, and the colonists generally, inveigh loudly against the supineness of the Governor, whose conduct may perchance admit of some excuse, though we confess ourselves unable to imagine that any can be offered for it. An efficient body of police to keep order at the diggings would be costly, and that is all. It can no longer be pleaded that there is not money enough in the colonial treasury for such a purpose. The licensed diggers at Mount Alexander pay in fees an average of £600 a day....

The 'well-disposed miners' had every reason for their fury. Many were well enough armed to look after themselves, and there were occasional outbreaks of 'Yankee Justice' against criminals. The majority of villains came from the convict settlements of Van Diemen's Land (Tasmania) some of them having escaped, others as ex-convicts or ticket-of-leave men, and these were known as Vandemonians. No doubt they got blamed for crimes committed by others, but they were certainly the hard core of the criminal element in Victoria. Even at Mount Alexander crime, never as wide-scale as alarmists alleged, could have been controlled by a moderately competent police force, but the diggers had to endure the help of men who were inefficient, corrupt, or brutal, some of them all three. And, as we shall see, they were led by a monster. It was to be the ever-continuing complaint against licence fees that was to lead to armed rebellion, but Authority and its minions were more to blame than any fee.

Not that the fee was a negligible one for a digger who was out of luck, as most were after the first heady strikes. Soon, men were averaging only £8 a month, and, inevitably, many of those who earned that or less tried avoiding payment. The police, who might well have been more worthily engaged in tracking down highwaymen or con-men who filled diggers with drink, then robbed them of their gold, took a great delight in hounding licence dodgers, or alleged ones.

The authorities had a problem. It was not easy getting good recruits for the police, and those who did join tended to be either inexperienced youths or hardened men more used to dealing with convicts than ordinary citizens. One man in the gold-fields service, Rudston Read, flatly stated that 'it was not in their nature to perform their duty quietly without bouncing, bullying and swearing at everyone', and if a respectable digger dared to object he

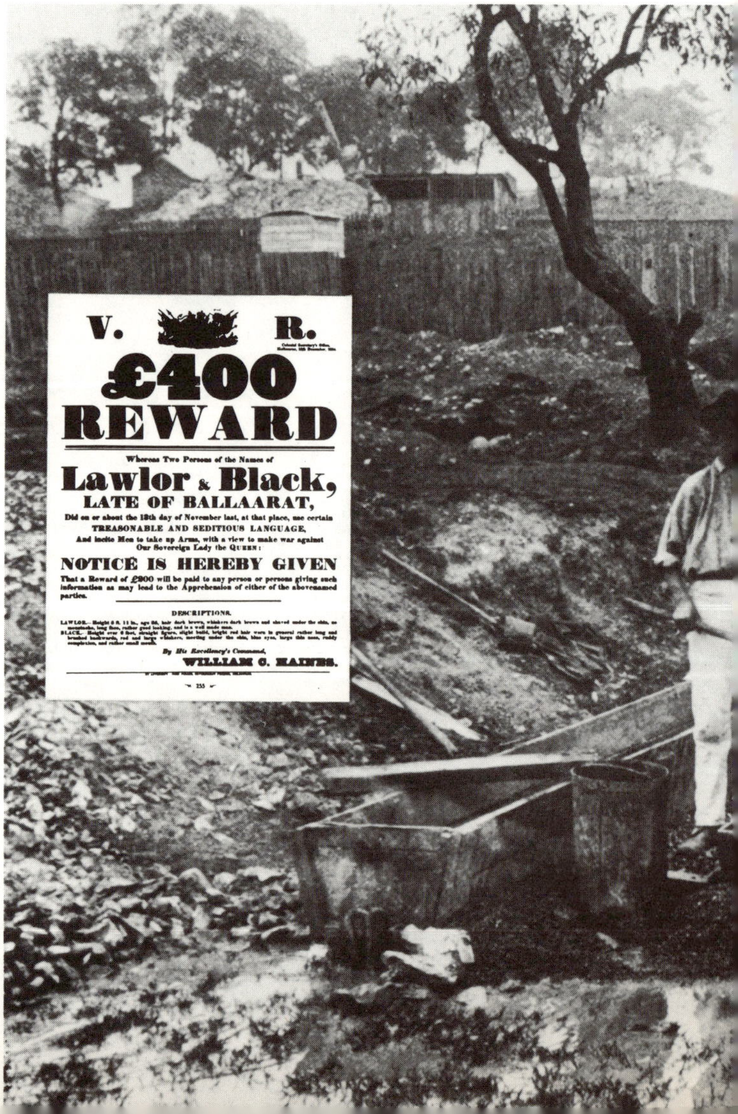

The poster was issued in the aftermath of the Eureka Stockade battle, 'Lawlor' actually being Peter Lalor. The picture is a more

was flung into handcuffs and charged with obstructing the police.

As for the honesty of these lawmen, Rudston Read declared that many policemen left the mines after several months the richer by £1,000. It was too easy to make money. There was prohibition at first on the gold-fields, but many of the police were in league with the owners of 'sly-grog-shops'. True, many of these shops were burnt down and their contents confiscated, but many more remained in business.

The miners might have forgiven their policemen for being on the take, if they had also been doing their duty stamping out crime and being the friends of ordinary, law-abiding citizens. After all, the police could not hope to strike it rich themselves. Instead the diggers were driven to hate all policemen, even the minority of honest ones, by the tactics of the majority. Soon, some policemen were stopping every man they saw and demanding to see his licence. An individual might be forced to show his six times in a day, and if he had left it in his tent, he was dragged to the Government Camp and was liable to be fined up to five pounds, or put in irons if he dared to object.

Even the Lieutenant Governor of Victoria, C. J. Latrobe, became uneasy at what he heard of the situation, but excused himself into believing that he had to make the best of what materials he possessed. One of those materials was Police Inspector Armstrong, the 'monster' referred to earlier. The diggers called ordinary police 'blood-hounds', but Armstrong was known as the 'Flying Demon'.

He appeared the very model of a policeman to Lord Robert Cecil and others. Cecil accompanied him on a raid on a sly-grog-shop at Ballarat in 1852. The Chief Commissioner for the Victorian gold-fields, Willian Henry Wright, was with them for part of the time (he was by no means unsympathetic to the miners, whom he considered over-taxed), but the actual raid was witnessed only by Cecil, for whom Armstrong seems to have been on his best behaviour, enough to quell any doubts that Cecil may have had about him. 'He is a very striking man,' Cecil wrote, 'well-made, tall, muscular, with keen "flashing" eyes, a splendidly clever countenance, perfect temper, and a quiet, fearless energy. His fame has gone far and wide through the diggings.'

Not until we meet 'Soapy' Smith in the Klondike, will we find another confidence trickster to equal Armstrong, whose real character has been left us by William Howitt, though he calls him 'Hermsprong'. Howitt's long and savage denunciation includes an account of the most notorious exploit for which Armstrong was remembered, though some of the policeman's feats with his whip (called 'Green Apples', and with a brass knob at the thick end of its stock), were almost as famous. He would knock down a man with it 'for half a word, or for a look only,' wrote Howitt.

A sympathetic Commissioner was the source of Howitt's information about Armstrong's masterpiece:

> A poor Irishwoman was left a widow with several children, the youngest of which was only a few days old. Hermsprong had discovered that this poor woman sold grog. He appeared before her tent, with his myrmidons, and, ill as she was, summoned her out. When he charged her with the sale of grog, she did not deny it, but said that her husband being killed by an accident, her countrymen had advised her, as her only means of support for herself and little children, to sell grog, promising to give her their custom; and the poor woman said, piteously, 'What, your honour, was I to do?'

peaceful Australian mining scene – and a reminder of the sheer hard work involved

> Without replying to her remark, Hermsprong turned to the police with him, and said, 'Fire that tent!'
>
> The poor woman shrieked out, 'For God's sake, sir, spare my tent! Spare my children!' The children were all at the moment in the tent; the infant of a few days old fast asleep. The police . . . refused to a man to execute his diabolical order. Swearing at them for what he called their 'd---d nicety', and threatening their dismissal, Hermsprong leapt from his horse, stalked up to a fire burning before the tent, seized a flaming brand, and fired the tent with his own hand.
>
> The poor woman, uttering a frantic cry, rushed into the tent, snatched up her baby, and, followed by her other children, came out, and stood shrieking and tearing her hair like a maniac, while her tent, and all that she had in the world, was consumed before her eyes.

Then the monster and his men rode away, with every digger in the area howling execrations at them.

For two years Armstrong's reign of terror continued until the stench of his methods could no longer be ignored, and he was dismissed. He retired laughing. Though his official salary had only been £400 a year, he boasted: 'I don't mind being turned out; for in these two years I have cleared £15,000!'

As alluvial gold became harder to find, so the miners' resentment at the licence fees grew. They practised equality and 'mateship', and their resentment was increased because of the power of privilege beyond the diggings, and the fact that they, the men who had brought colossal wealth to the country, had no say in the governing of it. A new Governor of Victoria, Sir Charles Hotham, arrived in mid-1854 to find discontent growing daily, until in October it finally boiled over.

Liquor licences had now been introduced, partly to raise more money, and on the night of October 6, a murder occurred at an inn. This was the Eureka Hotel, run by James Bentley, Mrs Bentley and John Farrell, an unsavoury trio who had been convicts in Van Dieman's Land before coming to Ballarat, where the hotel was situated.

The victim was a Scottish miner named James Scobie, and the trio were accused of his murder, only to be found not guilty. Convinced that the acquittal reeked of corruption, the diggers banded together, marched into town, and burnt down the hotel.

Hotham speedily sent 450 soldiers and police to Ballarat, while ordering a board of enquiry to investigate the charges of corruption, at which point a barman turned informer and the three were convicted of manslaughter. It also transpired that the stipendary magistrate, Mr Dewes, and a police officer had respectively 'obtained

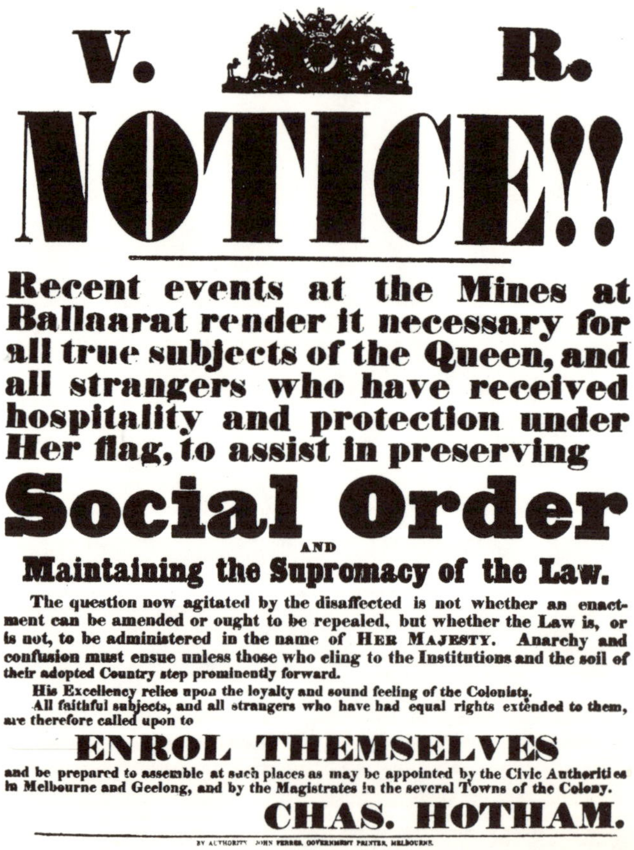

V. R.

NOTICE!!

Recent events at the Mines at Ballaarat render it necessary for all true subjects of the Queen, and all strangers who have received hospitality and protection under Her flag, to assist in preserving

Social Order

AND

Maintaining the Supromacy of the Law.

The question now agitated by the disaffected is not whether an enactment can be amended or ought to be repealed, but whether the Law is, or is not, to be administered in the name of HER MAJESTY. Anarchy and confusion must ensue unless those who cling to the Institutions and the soil of their adopted Country step prominently forward.

His Excellency relies upon the loyalty and sound feeling of the Colonists.

All faithful subjects, and all strangers who have had equal rights extended to them, are therefore called upon to

ENROL THEMSELVES

and be prepared to assemble at such places as may be appointed by the Civic Authorities in Melbourne and Geelong, and by the Magistrates in the several Towns of the Colony.

CHAS. HOTHAM.

BY AUTHORITY: JOHN FERRES, GOVERNMENT PRINTER, MELBOURNE.

loans of money' and 'taken bribes', and both were dismissed.

Instead of calming down, the miners began holding mass meetings, culminating with one at Bakery Hill on November 11, when they demanded–there were 10,000 present––proper representation, manhood suffrage, no property qualification for those on the legislative council, salaries for members, and short parliaments. It was very much like a Chartist manifesto, additional local demands being added: licences to be abolished and the local commissioners dismissed.

Hotham was not unsympathetic to the demand for the franchise, though he suspected foreign agitators caused the political unrest. Meanwhile, some Ballarat miners, particularly irked by the licence hunts because they now worked down deep shafts and wasted so much time waiting on the surface, were asked to wait still further while matters were sorted.

At once the situation got out of hand, with the local press and some professional agitators making things worse. Hotham promptly lost his head, sent up more soldiers, and ordered more licence hunts.

On November 29, another mass meeting, addressed by mob orators, who were certainly on the side of the angels, ended in licences being burnt on a bonfire. Hotham's absurd answer was another, more general licence hunt and, after a brief clash in which eight diggers were arrested, the reading of the riot act.

The next excitements were later summarised by Hotham in a dispatch to Sir George Grey, the Secretary of State in London:

> The aspect of affairs now became serious; the disaffected miners formed themselves into corps, elected their leaders, and commenced drilling; they possessed themselves of all the arms and ammunition which were within their reach, they established patrols, and placed parties on the high roads leading to Melbourne and Geelong; searched all carts and drays for weapons, coerced the well affected, issued orders, signed by the 'secretary to the commander-in-chief of diggers under arms', despatched emissaries to the other diggings to excite the miners, and held a meeting, whereat the Australian flag of independence was solemnly consecrated, and vows proffered for its defence.

The flag was the Southern Cross, with silver stars on a blue background, and the chief of the 'diggers under arms'–and the most admirable of the leaders–was Peter Lalor, whose oath is quoted at the head of this chapter. Before the oath, and holding his rifle in his hand, he said: 'It is my duty now to swear you in, and to take with you the oath to be faithful to the Southern Cross. Hear me with attention. The man who, after this solemn oath, does not stand by this standard, is a coward at heart. I order all persons who do not intend to take the oath, to leave at once . . .'

Lalor was born in Ireland in 1827, the son of a member of the House of Commons. A civil engineer, he had come to Australia in 1852. The rebellion historian was Raffaello Carboni, a fiery character who had fought for Italian freedom before becoming one of the leaders of 'The Republic of Victoria'. He was a true revolutionary. The other leaders were the Prussian Edward Thonen, the Hanoverian, Frederick Vern, and, less officially, an American named James McGill.

A stockade had been made from logs and slabs of mining and building timber that the men found lying nearby. It was sited at Eureka, a mile and a half along the road to Melbourne, and the force manning it rapidly reached 500, Lalor seeing to it that the many foreigners were split up among the various divisions, so that, as Carboni wrote, 'there was not one single division distinguished by nationality or religion.' It was in this stockade, whose construction was supervised by Vern, that the Southern Cross was hoisted, and drilling took place inside. The fortress was to prove less strong than it looked.

Most of the Americans on the gold-fields had the sense to remain sympathetically neutral to avoid being branded as instigators. They formed themselves into the California Volunteers, ready to act if necessary, but remaining independent. Meanwhile, the diggers failed to establish security in their fortress, allowing anyone in, including government spies. The authorities circulated a story that cannon were on their way to Eureka, and McGill led out some of the best-armed diggers, fatally weakening the stockade. And the rebels were further weakened by diggers who became disenchanted or just plain scared.

On the night of December 2, 1854, the diggers waited anxiously for news of McGill's expected victory, but the next morning the government forces attacked with, according to the official records, 30 cavalry, 70 mounted police, 152 foot soldiers, and 24 foot police, in all a total of 276 men. Legend has built it into a far bigger army.

One of the best accounts of the action was written by C. D. Ferguson, a California Ranger who had been accused of cowardice for advising caution and stating flatly that the diggers would fail. In the event, he and some of his fellow-countrymen were in the stockade when it mattered, as they had planned to be.

It was 3 a.m., the troops having planned to arrive when many diggers were sleeping in their tents and shacks away from the stockade:

> They had come down on us just as the light of day was

breaking in the east. We were formed in line, and the first order received was 'California Rangers to the front!' The Fortieth regiment was advancing, but had not as yet discharged a shot. We could now see plainly the officer and hear his orders, when one of our men, Captain Burnette, stepped a little in front, elevated his rifle, took aim and fired. The officer fell. Captain Wise was his name.

This was the first shot in the Ballarat war. It was said by many that the soldiers fired the first shot, but that is not true, as is well known to many....

No sooner did [Captain Wise] fall than the soldiers were ordered to fire on us, which they did, and then charged. The fire had a terrible effect, but we returned it with like effect, as deadly as theirs. Just at this time, when the splinters from the timbers of the breastwork were flying the thickest, Vern came running past. I asked him what he was running for. 'To stop the others,' was his reply. I had my own opinion about it.

It was now the most exciting time I had ever witnessed. It was a hand-to-hand fight. The soldiers were in among us. Lalor was shot in the arm, and Hall pulled off his necktie and we wound that around it. He was bleeding profusely and before we were through had fainted from loss of blood. We put him in a shallow hole and covered it over with some slabs.... I was near poor Ross, and he said, 'Charlie, it is no use, the men have all left us,' and the next instant he said, 'My God, I am shot,' and fell. Before I had time to look and see how badly he was hurt, a soldier demanded my surrender, to which I politely answered that I would see him dam'd first, and made my first attempt to escape. In the excitement I had not missed the rest, and upon looking around discovered that I was almost alone....

As I jumped the stockade I fell, and the soldier who had demanded my surrender fired, and the ball passed through my hat. The fall resulted in making me a prisoner. I was not long, however, in getting onto my feet, but found a party of troopers had headed me off in that direction. Turning I jumped back into the stockade, but was there met by any number of soldiers. I attempted to rush through, but was seized upon by several and we had a rough and tumble for a few brief seconds, and I finally got through and struck for another place to make my escape. The soldiers had been ordered to cease firing, but the police kept it up when they saw a poor fellow trying his best to get away....

The resourceful Ferguson managed to surrender, not to the soldiers who now had him cornered, but to Captain Carter of the police. Another American jumped onto a rope and slid down a hole 100 feet deep, later climbing out to safety, Around 25 diggers had been killed and others died of their wounds, while three privates and one officer died, and a number were badly wounded. The battle had lasted at the most some 20 minutes.

The immediate aftermath of the battle saw terrified prisoners expecting to be hanged (their captors told them they would be), and a certain, if disputed, amount of police brutality. But the defeated diggers, for all that they represented the minority of miners who were prepared to add action to mere complaining, undoubtedly won a moral victory. Fifteen were tried for treason and acquitted, a Gold-fields Commission was appointed, and a general amnesty proclaimed. The hated licences were abolished and replaced by an export duty of 2s 6d an ounce on gold, while miners were given official rights on payment of £1 a year.

Lalor lost his wounded arm and hid with a price on his head, but this was later withdrawn, and, as befitted the hero of the rebellion, he became enormously popular, later becoming a conservative member of the Victorian parliament and finished up as its Speaker. He refused a knighthood.

Vern, as unsavoury as Lalor was attractive, having run away during the battle he had done so much to promote, lost his few remaining friends in the months ahead by his behaviour, but survived. Carboni wrote his invaluable account of Eureka, but vanished into obscurity. Ferguson, who was later to resign from the Burke and Wills expedition into central Australia on the grounds that it was incompetent, extricated himself from his post-Eureka problems with considerable skill, being discharged at his trial after giving the impression that he had been a mere innocent onlooker. Later, he met Governor Hotham, who questioned him closely about the grievances of the diggers. Hotham died in 1855, a by-no-means unsympathetic character, who was accused by the British government in the person of the Colonial Secretary, Lord John Russell, of bungling the case for the prosecution of the rebels. He deserved better of his country.

As for Eureka Stockade, it gradually became a symbol with an importance far in excess of its immediate and essentially local significance. As Henry Lawson put it in his *Eureka*:

> But not in vain those diggers died. Their comrades may rejoice
> For o'er the voice of tyranny is heard the people's voice;
> It says: 'Reform your rotten laws, the diggers' wrongs make right,
> Or else with them, our brothers now, we'll gather to the fight.'

A scene from the 1947 Anglo-Australian film, *Eureka Stockade*

New Zealand and Back

I shovelled away about two and a half feet of gravel, arrived at a beautiful soft slate and saw gold shining like the stars in Orion on a dark frosty night.
Gabriel Read on his strike in Otago, 1861

With the hated licence tax out of the way, the miners –first in Victoria, then elsewhere–found a new enemy, the Chinese, who began arriving in vast numbers from 1855. By 1861, there were some 204,000 Europeans on the Victorian gold-fields and 24,000 Chinese, only six of them women. The alleged sexual depravity of the newcomers has been noted earlier, but some of the hostility towards them was sheer racialism, the forerunner of the White Australia policy.

Yet, undoubtedly, the main objection to the Chinese was the lessening yield of gold, with the prospect of their becoming a cheap labour force to mine what was left. Times, indeed, were changing. Though new strikes did occur, it was becoming steadily harder to find alluvial gold. Deep shafts and expensive modern machinery were taking over from the pan and the rocker on the older fields, and large companies were busily crushing quartz. So a flood of cut-price Chinese labourers was a potential threat.

New South Wales and Victoria tried to discourage the unwanted immigrants by levying a stiff poll tax of £10 and limiting the number of passengers per ship, but the Chinese simply landed in South Australia and walked up to the diggings. In 1857, on the Buckland River, the whites rioted against the newcomers, after which a tax of £6 a year was levied on each Chinese, while South Australia tightened up immigration restrictions. But worse rioting was to follow, this time at Lambing Flat in New South Wales in 1861.

After initially complaining that the Chinese spoiled the water, were dirty in their habits, took up more ground than they were entitled to etc., the diggers began to take action–including a 'processional march of some 6,000 diggers, armed with pick handles, revolvers, bowie knives, &c., and headed by a brass band and banners,' as George Preshaw, a banker at Burrangong (Lambing Flat), later wrote. The object of the procession was to frighten local storekeepers who supplied Chinese and burn their stores if they continued to do so. The climax of the occasion came when the mob stormed the Great Eastern hotel hoping to lynch the special correspondent of the *Sydney Morning Herald* who had dared to criticise them for disloyalty to the Government. When they found he was not there–he and a policeman were next door hiding in the Oriental Bank–they fired a volley through the hotel's roof and went on a wild drinking spree, the landlord supplying plenty of free booze to keep the mob pacified.

Things got steadily worse, though Preshaw insisted that the majority of the mob were not miners at all. One of the worst outbreaks occurred on a Sunday, when the 'monsters on horseback', as the *Herald's* correspondent called them, attacked a hill where some 300 Chinese were located. The encampment was ruined and burnt, a number of Chinese had their pigtails pulled out by the roots and placed, like scalps, on the diggers' banners, and any gold found was stolen. Then the band played *Rule Britannia* and the diggers proceeded to another Chinese camp, this time at Back Creek, where there were 500 unfortunates.

Incidents there included an attack on a woman married to a Chinese, the mob setting fire to her child's cradle, and a discussion as to how to gang rape her, which mercifully was foiled by the intervention of less brutish whites. Finally, troops and police arrived to quell the riots and a pitched battle took place after the Riot Act had been read. Only one man was killed–a digger–though a number on each side were wounded. A dark night helped keep casualties down.

A number of rioters were arrested, but all were acquitted by sympathetic juries, which caused the *Sydney Morning Herald* to comment that it was 'a fitting wind-up for so disgraceful a commencement with which it is so perfectly in keeping.'

While these riots were disrupting life on the Australian gold-fields, New Zealand was experiencing the heady, sudden joys of a rush. Not that gold played so decisive a part in the history of the two islands as it did in Australia, for the Maori wars and sheep, especially the latter in the long term, were the key factors in mid-Victorian New Zealand. Nevertheless Dunedin and Christchurch were transformed, Central Otago was settled and inroads were made into the forests to the west, and, as always, miners brought their own form of instant democracy while trade and finance flourished.

As early as 1852, money was being offered in Auckland for anyone finding gold, but for some years only small finds were made. The first big strike, which earned its finder £500 from the Provincial Council of Otago, was made in 1861, either on May 25 or 26, by Gabriel Read, who had prospected in California and Australia.

He 'struck colour' on the bar of a stream in the district of Tuapeka, west of Dunedin, and reported to the Superintendent on June 4, stating that he had panned seven ounces for ten hours' work. After checking his report, the news was made public and Dunedin promptly emptied in the by-now traditional manner. The city, of course, rapidly filled again as Victorians poured into the port and business boomed. The population doubled to 26,000 by October.

Profiting by the earlier rushes in Australia, the

Government assumed responsibility in an efficient manner from the start, doing its best to meet the diggers' demands for better roads, coal supplies, a hospital and properly organised townships. The first season resulted in some 400,000 ounces of gold and remarkably little crime. But there was also remarkably little in the way of comforts on the gold-fields, especially in the winter of 1862 (July and August) which was a severe one, rugged enough to send some of the Victorians back across the Tasman Sea.

Meanwhile, other diggers were heading west, and two of them, Hartley and Reilly, went up the Molyneux River and found clear signs of a rich field in the Dunstan Mountains. they later returned to Dunedin and divulged their secret after being promised a £2,000 reward. A rush began, not only from the city, but from the Tuapeka field, where 5,000 left in a week.

This find and a later one near Lake Wakatipu were in striking, but wild country, a land of mountains and gorges, of forests and places so short of timber that rockers could not be made or claims pegged out, while the climate veered from hot summers to bitter winter frost and snows. At one small camp on the Arrow River a miner named Fox not only elected himself Commissioner and allocated every miner 60 feet along the river, but announced that he would fight any man that jumped a claim. He and his willing subordinates found 232 pounds of gold before other miners appeared on the scene.

It became harder for the authorities and the police to regulate the miners as they headed into the wilds, and harder to organise escorts for the gold. Major incidents, however, were few, such as when 'a mob of Tipperary men were going about and jumping portions of the richest claims' at the Arrow River diggings. A few Chinese appeared after sending scouts over to find out if pigtail hunting was likely to be a local sport, while mining gradually became more organised. Even today, central Otago is a gold-producing area via expensive machinery, the wilder country being a fruit-growing area and a holidaymakers' paradise.

There were also major finds on the west coast of South Island, the main town being Hokitika, which had a typically instant birth in 1864, and was soon booming. Most people arrived by sea and had to cross a river bar, and many a ship was hurled onto the beach by the surf. Inland, it was easiest to keep to the river and creeks, and hazards included winter rains and flooded claims, dense forests, mosquitoes and roads with 'roots of all sizes, torn and mangled when small into a sort of macaroni squash, and when large remaining a dead hindrance to both horses and men [they] caused the mud ploughed by cattle and pack-horses to assume the appearance of a torrent; so bad was it that the whole distance was marked by the bones of

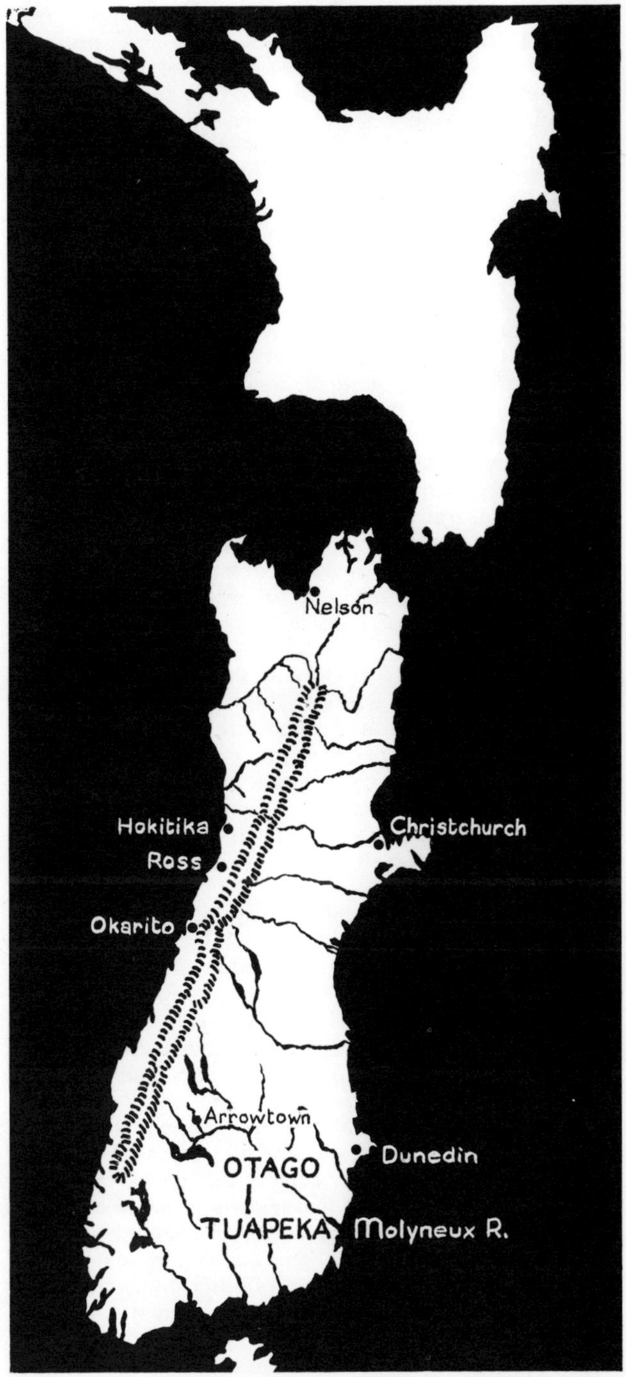

Gold Rush, New Zealand

dead animals.'

By 1866, there may have been as many as 30,000 people on the west coast, with new rushes occurring further north than previously near the Buller River, and the following year gold was found on the Coromandel peninsula of North Island. Not until the 1890s, however, long after the era of the gold rushes, did mining again boom, on the North Island at the Waihi mine and on South Island not far from where Hartley and Reilly originally struck it rich.

By the 1890s, the Western Australia gold-fields were world famous, but one area 2,000 miles from the fabulous finds there must be visited first: Queensland. There, gold was found in 1858, but the first great strike was made in 1867 by James Nash, who announced his find to the Commissioner for Public Lands in a letter dated 'Brisbane, April 27, 1868'.

It happened on the upper waters of the Mary River, where Nash began prospecting in September. He managed to keep his secret until he had made sure of his reward, then the rush began, and by March, 1868, Nashville, later renamed Gympie, had a main street half a mile long, and a rip-roaring line in Saturday nights: E. B. Kennedy described them in *Four Years in Queensland*:

> *The* night *par excellence* was Saturday night: the whole length of the street was so full of diggers that we could hardly move at all, and what with singing, swearing, fighting, drinking, bargaining for loaves, beef and sausages for Sunday's dinner, the noise was tremendous, while every public-house was crammed with men discussing their various finds . . . while they frequently paid for their drinks with small samples of gold.

Claims at Gympie were 40 feet square per digger, and holders had their section of river bank to which most of the pay-dirt had to be carried from a mile or so away in carts or barrows. There were some sensational finds, including one 1,000-ounce nugget, and once again the Chinese began to flood the gold-fields, even outnumbering the whites in Queensland in the 1870s. Discriminatory legislation was passed against them, climaxed in 1878 by a ruling that they should be excluded from any gold-field for three years after its discovery. This ban was also applied to other Asiatics, and to Africans as well.

Most remarkable of all Queensland mines was Mount Morgan, named for three brothers who bought a hill-top for £1 an acre from an owner who had no idea of its worth. The Morgans built a crushing mill and found that their hill, which had cost them £640, contained crumbling ironstone in which was gold of such purity that it produced 30-40 ounces per ton, paying £1 million in a single year in dividends. Over 25 years this fabulous square mile produced £14 million.

For the record, and before finishing the Australian golden saga in Western Australia, it was New South Wales which saw the largest mass of gold ever mined in one piece. The place was Hill End, 30 miles north-west of Bathurst, the date was 1872 and the finders were partners called Holtermann and Beyer. The giant nugget weighed 630 lbs, a huge lump of black slate and gold, the latter being worth around £12,000.

Two decades before this fabulous find, the first reward was offered to anyone who could find gold in Western Australia, and no less a person than Edward Hargraves tried his luck in 1862, without success. A certain amount was found in the Kimberley district in the late 1880s, and then a boy named Jimmy Withnell triggered off a rush in the Pilbara region 70 miles east of Roebourne.

The Holtermann nugget was the biggest mass of gold ever mined in one piece, though it was a mixture of black slate and gold. It was worth £12,000 in 1872

The 15-year-old youth and his brother had been splitting slabs to build a hut on their sheep station, and returned to their tucker-bags to find crows enjoying the contents. Jimmy reached for a stone, which had a yellow mark on it that did not rub off. When it was examined in Roebourne the excitement was colossal and the Government Resident hastily sent off a telegram to the Colonial Secretary, which began 'Jimmy Withnell picked up a stone to shy at a crow...'

Enflamed by gold fever, the Resident failed to complete the message, which was greeted with the following reply: 'What happened to the crow?'

The date was Christmas 1887, and the north-west of the immense colony, which at that time only contained 40,000 people, had received a welcome Christmas box.

Scarcity of water made rockers and sluice-boxes impractical in Western Australia so a process known as 'dryblowing' was used. First a 'paddock' had to be stripped of 'all overburden', in other words stones were taken off the chosen spot and lumps of clay and earth were broken up. Then the digger could dry-blow by hand or with a machine. The dryblower was a four-legged affair with a square hopper on top onto which the dirt was put; then the dryblower was shaken. While any nuggets would remain on the hopper, the finer dirt went through the holes onto a tray, placed at 45 degrees. The tray has 12 riffles (called ripples in Australia) and the bottom of this tray was perforated, allowing an air blast from bellows rigged below to force dust from the stream of gravel coming down from the hopper.

While the sieved and blown gravel passed over the tray, any gold was caught by the riffles, and when they were full, the tray was removed and its contents put into one of two dishes. The digger then stood sideways to the breeze, held the loaded dish as high as he could, then tilted it gently to let its contents fall into the second dish at his feet. The breeze blew away the ordinary dust and what was left of the lighter gravel, while any heavier matter, including gold, fell into the other dish. Then the dishes were reversed and the sequence continued until no clogging dust was left.

Now the digger went to work with the dish held knee high, shaking and swirling and gradually lightening what was left, finally using a rotary motion to get the last dirt onto the side, from which it was blown away. The remains–if luck was in–would be gold. It was very hard work, and so was the simpler process, which cut out the machine and concentrated on working the two dishes. Fortunately the alluvial was not too deep, but another problem could be moist ground, despite the lack of actual water. Sticky soil was sometimes treated with a sieve, but when the ground was really moist, fires were lit and shovelled onto the alluvial soil, so as not to hold up the dryblowing the next day. The process is yet another reminder of the sheer hard labour involved in finding gold, except for the lucky few.

The stage was now set for the final Australian gold rushes. By this time gold had been found in every part of the continent except South Australia, for even the Northern Territory had produced gold. That was in the '70s, but the cruel climate and the sheer difficulty of getting to and from the strikes, lessened the importance of these finds. The only really successful diggers in the north were the despised Chinese, and many whites died or had their health destroyed. Mrs Dominic Daly, whose husband had been a governor of South Australia, knew the Northern Territory at first hand. She wrote vividly of the track between Darwin and Kimberley:

> All along the track... the remains and relics of white men were found. Here a scrap of clothing... there a strap with the owner's name; pack saddles torn and plundered, empty meat tins, and initials cut on trees where the poor gold-seekers had camped for a night or so; an old hat and the bowl of a pipe, were the only traces left of men whose fate will always remain a mystery.

Kimberley was far nearer Darwin than to the spot where Messrs Bayley and Ford, old Queensland hands, made history one Sunday afternoon in August, 1892, striking a rock with a tomahawk at a spot whose name rang round the world–Coolgardie. It was some 100 miles east of Southern Cross, where Bayley returned in September with 500 ounces and promptly emptied the settlement, where miners earning wages were on strike because their pay had been reduced.

The first 700 men to reach the golden spot had one main problem–water. The Hon. David Carnegie, son of the Earl of Southesk, was there to describe how it was brought to the camp of tents amid the eucalyptus trees:

> [It] was carted by horse teams in waggons with large tanks on board, or by camel caravans, from a distance of 36 miles, drawn from a well near a large granite rock. The supply was daily failing, and washing was out of the question; enough to drink was all one thought of; two lines of eager men on either side of the track could daily be seen waiting for these water-carts. What a wild rush ensued when they were sighted! In a moment they were surrounded and taken by storm, men swarming onto them like an army of ants. As a rule, eager as we were for water, a sort of order prevailed, and every man got his gallon water-bag filled until the supply was exhausted.

There was alluvial gold near quartz reefs, which caused some problems, but water remained the chief one until the

summer drought ended and thousands more diggers poured into Coolgardie, which rapidly became the liveliest spot on the continent, and from which prospectors headed out to seek new fields nearby. Far the most important was the rich reef found by two miners from Adelaide, members of a syndicate, a month after three prospectors had struck alluvial gold in the area, 24 miles north-east of Coolgardie. The three, Hannan, O'Shea and Flanagan, put in for their reward claim in June, 1893, while the reef–the future 'Golden Mile' of Kalgoorlie–was struck in July, though not until 1896 was its extraordinary richness fully realised.

Kalgoorlie, Coolgardie and other Western Australian mines were to produce gold worth £22,200,000 between 1892-1900, and when *Back to the Gold-fields* celebrations were held in 1938, Kalgoorlie alone had produced more than £100 million. Western Australia was destined to become the leading gold-producing state in Australia, producing 80% of the nation's gold.

Prospects had not seemed so bright back in 1894, when the inhabitants of Coolgardie began to fear a depression. And the Londonderry Golden Hole discovery ended in fiasco.

It started brightly enough, when a party of six set out from Coolgardie early in 1894. They had no luck, and were heading back to look for employment at day wages, all of them travelling separately now and hoping to strike lucky at the last minute.

John Mills, born in Londonderry, and lately at the New South Wales mines, settled down with his pipe one evening to think about his lack of success. John Marshall, Scottish born, and the secretary of the Coolgardie Gold Diggers' Committee, described what happened next:

> The huge reef which he had tested in many places, without seeing a single speck of gold, lay at his feet. Sitting thinking ... he almost almost unconsciously rubbed his heal against the huge, moss-covered outcrop at his foot, and carelessly looking down he suddenly caught the glint of some shiny substance on the rock. Lazily raising himself he ... found to his intense astonishment and delight that it was a piece of stone full of gold. Breaking several pieces of the cap of the reef, he found to his amazement that *it was literally hanging together with gold*!

He showed his mates the treasure spot, and they 'toiled like galley slaves' for weeks 'with the rudest appliances'. The work was deadly dull and extremely tough, but they found some 8,000 ounces of gold, worth £32,000. Then Mills and one of the others went back to Coolgardie with suitably long faces to apply for a lease of 24 acres that might possibly contain a little gold. A mining agent named

Once a prospector, always a prospector! A minority went on all their lives seeking the elusive gold

Lindsay went out to the fabulous spot, and succeeded, and meanwhile one of the older members of the gold-rich six got ill and returned to town, where he went on a drunken spree and let the secret out. At once Coolgardie emptied, the 'Londonderry' was reached, and Mills showed off some choice specimens. Otherwise, there were no signs of gold, but a company was floated in London and Paris after the Resident on the field, Lord Fingall, had bought the property. And the 'Golden Hole' was sealed over with a strong plate.

Alas, when it was reopened some months later, after the company was in business, 'it was found, to the utter amazement and dismay of all concerned, that the kernel had been taken and only the worthless shell was left.'

Considering the vast distances involved and the number of small gold-fields, Western Australia's rush was comparatively peaceful. The diggers used a non-lethal version of Californian instant justice, a tin pan being beaten like a drum if a man was caught stealing, at which all the miners converged to the spot. The culprit was tried on the spot and if he had been caught in the act, he was banished and told never to return on pain of tarring and feathering. If police were present on the gold-fields there was rarely trouble, and nor was there much need to escort gold so strongly as in the eastern states, for a robber could hardly get far in the endless wastes except by horse or camel, and they could be followed, while the few spots with known water could be alerted. On the most isolated fields, a load was often entrusted to a single man. A greater enemy than any bushranger was the bush itself. Aborigines, admittedly sometimes hostile but almost always ill-treated across the continent, proved invaluable guides to diggers, only a minority of whom became true bushmen.

Western Australia proved an invaluable place for those wanting to 'disappear' from sight for a while. As one contemporary ballad put it:

But though we plumbed the depths of many mysteries and myths,
The worst we had to fathom was the prevalence of Smiths.

The 'Smiths' ranged from criminals to errant husbands.

One day in 1895 Coolgardie exploded with excitement when the local paper broke the news of a sensational find of alluvial gold. Rumour was rampant, as groups of diggers set off in every direction to the alleged spots–there were plenty of them–where the gold was said to be. They headed into the bush on foot, on horses, camels and bicycles, hundreds of men who should have had the sense to wait like the prudent minority before heading they knew not where.

One paper named a digger called McCann as the man who had made the strike, but the question being asked was a basic one–where was he? Soon thousands of angry miners were pouring back into town their tempers frayed. They had pertinent remarks to make about the paper for publishing false information, and the paper blamed McCann.

John Marshall, still the secretary of the Gold Diggers' Committee, was coming down Bayley Street when he saw an excited crowd trying to break into the offices of the *Miner*. He had been worried enough about diggers getting lost in the bush, but this mob outbreak was more urgent. Then he saw something which made it even more dangerous:

A number of the baser fellows ... were roughly using a

The interior of a gold quartz-crushing battery on the Thames gold-fields in New Zealand tall man, who appeared to be much agitated, and whom I learned was none other than McCann himself... Now, I thought, is the time to act, if at all. I therefore raised my voice to its highest pitch, and cried, 'Look here, boys! You know that I am John Marshall... I am determined to take steps to locate the alleged "rush", and find out whether McCann's report be true or not...

He went on to say that he would organise a public meeting to discuss the matter, and that McCann would be made to lead a party to the place he was supposed to have found. Marshall himself would be in charge of him.

An immense crowd turned up at the meeting, and when McCann, big, brawny, heavily moustached and deadly pale, stepped into the waggon which was serving as a rostrum, he was greeted with hisses and derisive cheers. He had taken the precaution of fortifying himself with plenty of drink, which was just as well, with so many below him believing that he was guilty of the greatest of all crimes, starting a bogus rush.

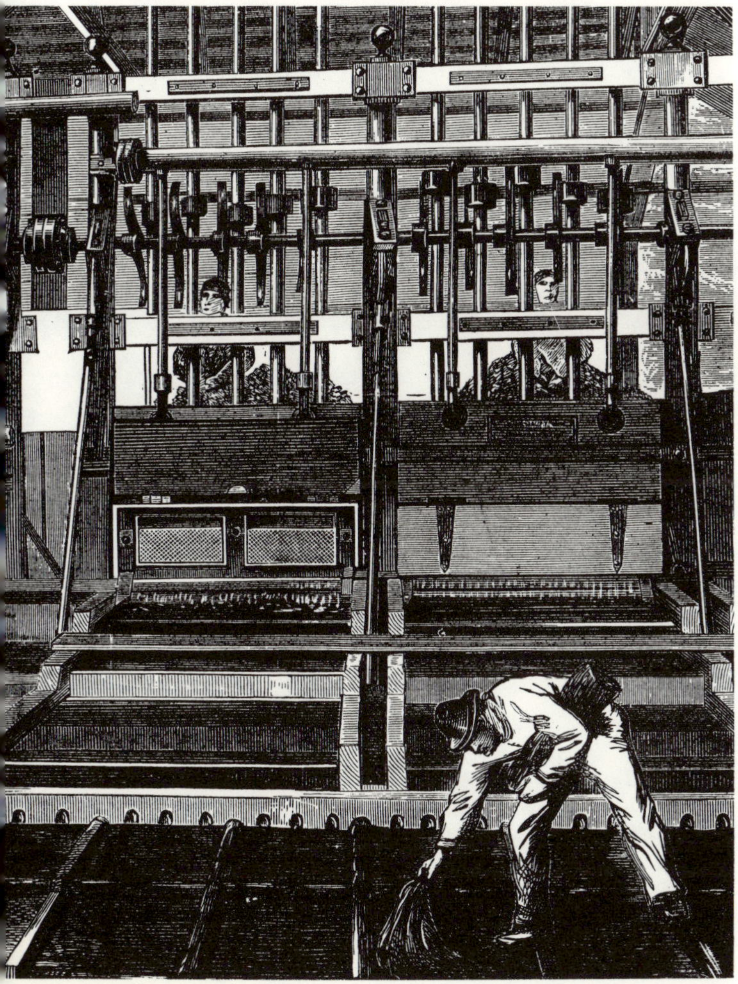

The more he spoke the more confident he became until he had many of his listeners believing him:

He actually volunteered to lead a party to the place where the alluvial gold had been obtained! His utterances were greeted with rapturous applause... It was proposed and carried that a party of four be sent out in charge of McCann ... and in a short time sufficient funds were raised to send away a party.

Marshall took him home for the night, only to find when he went to check that McCann had vanished. Fortunately, he reappeared in the morning looking miserable in the extreme.

The party left town, armed to the teeth, and with orders to bring back McCann dead or alive. Marshall stayed behind fearing the worst and hoping that the men would at least return on a Sunday when the pubs would be closed.

McCann came back alive on the Sunday evening. Not only had no gold been found, but it was all his four keepers could do to prevent lynching parties stringing him up on the way back. McCann had kept pleading with them for a revolver to end it all, but the four had no intention of putting him out of his misery. They got him to Marshall's house, and the latter rushed out the back to the police camp, some 50 yards away, the entire force at once turning out to guard the wretched man.

Meanwhile, a crowd rapidly gathered, and McCann bravely offered to face them, until persuaded by the police that the local lock-up might be a better spot for him.

Marshall described what happened next:

By this time the crowd was the largest I had ever seen on the gold-fields. When the report ... was read out showing that the alleged great gold discovery was a cruel hoax ... that thousands of men had been fooled, then thousands of pounds spent, and the lives of hundreds of men endangered through the senseless babblings of a drunken fool [there] occurred a scene which positively baffles description. Cries of 'Bring McCann out!' 'String him up!' 'Pull his liver out!' 'Tear him from limb to limb!' were wildly indulged in.

One of the mob shouted: 'Marshall's hiding McCann; pull his place down about his ears!' and it looked as if fury would soon turn to madness. So Marshall boldly stepped

onto a rostrum and faced the sea of hate-filled diggers. He waited till they fell silent and narrated what steps he had taken, imploring them not to 'be stirred up to deeds of violence by the pimps, parasites, and "spielers" ... I laid particular stress on the warden's warning, given to me, that, in the event of anything happening to McCann, we, who had assumed the responsibility of sending him away in charge of an armed party, would be held liable.'

Just as his eloquence seemed to be swaying the crowd, a man urged all present to storm one of the newspaper offices, and a rush began in that direction, only stopped by the police and a large number of law-abiding citizens. The mob broke up, muttering angrily.

The next morning McCann came and thanked Marshall for saving his life, then he disappeared. There was a grim, symbolic aftermath, for a few days later a horde of diggers poured into town from their camp where they had been waiting for definite news. Many had camels, and on the back of the leading animal a gallows had been erected bearing a full-size effigy of the hated McCann. Many joined the procession, which finally reached the offending newspaper office that had first published the story.

A bonfire was lit and the effigy was burnt to a chorus of howls, yells and roars of delight, and before the flames had consumed it, some of the mob tried to hurl it through the windows of the office so that it, too, would be set on fire. Fortunately for Coolgardie, a strong contingent of police were on hand and the diggers dispersed. The great McCann rush was over.

As a postscript to this example of diggers' justice, a miner named Winter cheated a friend out of £80 worth of gold near Kalgoorlie and was lucky not to be hanged. Instead, something almost as bad happened to him. Not only was he expelled from his camp: details of his 'trial' were sent to gold-field newspapers all over the State, which he left after being hounded from field to field. No doubt he was equally unwelcome outside its boundaries.

During the 1890s, a water pipeline was built to pump fresh water from Perth to Coolgardie and Kalgoorlie, a distance of more than 300 miles. This was more than a major feat of engineering, it was a sign of the times, which were indeed changing: soon after the turn of the century, an actress named May Vivienne visited Coolgardie and later wrote a delightful account of the place:

> Taking my bicycle I went for a tour of inspection around the various streets adjacent to the town, where I found many very nice houses, and to my surprise saw a lady in a very nice carriage drawn by a pair of greys. Truly, I ought to be surprised at nothing in wonderful Coolgardie. The roads here are the most level and the best for cycling I have ever ridden on ... Riding merrily along I was overtaken by a man cyclist, who did not favour me with more than a passing glance, lady cyclists being no rarity here.

Clearly, things had quietened down dramatically. Railways had come to both Coolgardie and Kalgoorlie by 1896, but lady cyclists ...! The gold rush period was now over.

The diggers did not open up Australia as much as the Argonauts and their successors opened up California and the American West. In many gold-bearing parts of Australia the land-owners and squatters were there before the rushes began.

So what did the Australian gold rushes achieve, apart

The Tuapeka goldfields in Otago, New Zealand

from the obvious financial gains to some diggers and many companies and investors at home and abroad, especially in Britain?

They notably changed a pastoral, authoritarian society into a democratic one, and because the diggers spanned the continent they helped mould the different colonies into one nation. Their individuality, initiative and sense of 'mateship' were to become part of the national character, which was to be seen at its staunchest and boldest at Gallipoli, on the Western Front, and in a later World War.

Yet for all their sturdy independence, the diggers had a basic respect for law and administration on the British model, which makes for less lively reading, but can hardly be criticised for that. Technically, the gold-fields developed more slowly than their American counterparts, the diggers being more suspicious of capitalists than their opposite numbers in the American West. The Labour Party's notable early triumphs in Australia (compared with its British equivalent) surely owed something to the digger spirit.

The dark side of the picture has been noted, the extreme dislike of the Chinese, and to this must be added the view of the Aborigines as Stone Age men, to be no more regarded than kangaroos. But it is pointless upbraiding these Anglo-Australians of long ago for attitudes which remain to this day in Britain and Australia with far less excuse.

In short, the diggers were a hard-working, colourful, all-too-human band of men who well deserve their legendary fame.

Bonanzas West

To see Virginia City and Carson since I first heard their fame in New York had been with me a passion. . . .
Charles Dilke: *Greater Britain*

The very month that Edward Hargraves felt himself 'surrounded by gold' in New South Wales, the first major exodus from the Californian mines occurred, spurred on by a strike by James Cluggage a few weeks earlier in January, 1851. He and his partners, John R. Pool, had been prospecting in Oregon's Rogue River Valley, and Cluggage struck it rich on Jackson Creek. They were soon averaging 100 ounces of dust and nuggets a day.

Oregon, which had lost so many settlers to the Californian mines, now saw many return, together with hundreds of strangers, and soon the boom town of Jacksonville was flourishing. Despite the success of this area, Oregon was not destined to be one of the great gold-mining states, and is best remembered at this period for some particularly depressing Indian wars in which atrocities by white 'volunteers' made Indian murderers seem like amateurs. 'The Cry was extermination of all the Indians by the whites and the company began to break up into small companies to go to different Indian Rancharies to clean them out,' wrote James Cardwell.

Some miners took time off to run stores. Herman and Charley Reinhart, German-born, but bred in New York and Illinois, set up a bakery-cum-saloon at Browntown in 1853, opening on April 10 to the strains of two violinists at $4 a day each plus board. The place never closed and takings ranged from $80 to $200 a day. Credit was allowed to miners for drink, cigars and pies. As well as plenty of food, gin, whisky and brandy were provided. So were 'French cordials', champagne and other 'fancy drinks', but they had to be brought in by pack mule from Portland, and Scottsburg and the prices for them were steep. The brothers had six private rooms for gambling and a hall for miners to play cards for food, drink and cigars.

The town also boasted two 'Dance or Fancy houses', several butchers shops, blacksmiths and other establishments, and the brothers later opened a bowling alley at a spot called Sucker Creek. But gold kept beckoning the pair away from the business, and in 1858 Herman headed for British Columbia. Jackson Creek remained the chief Oregon goldfield for many years. Browntown no longer exists.

Gold had been found in small quantities in British Columbia from 1850 onwards, but it was not until 1857 that strikes on the Fraser River and its tributary, the Thompson, led to a rush the following year, giving the local authorities immense problems from which they emerged with great credit.

This was still the domain of the Hudson's Bay Company, more than 1,000 miles from the Red River Settlement in what is now Manitoba. After the dispute over the ownership of the Oregon country had been finally resolved in 1856, making the 49th Parallel the boundary line to the sea, the small, isolated community in British Columbia began to prosper quietly. Its headquarters were at Fort Victoria on Vancouver Island, and agriculture provided enough food for the Hudson's Bay posts in the area. The local Indians, with European tools and ship's paint, developed their totem poles into a striking form of art. Each year a Company ship arrived, and naval vessels sometimes called to relieve the loneliness of the trappers and settlers.

When news of the finds on the Fraser broke, Victoria was still a village with a population of some 800. James Douglas, a remarkable man who combined the posts of Governor of Vancouver Island and Chief Factor of the Company in the region, shipped 800 ounces of gold, much of it found by Indians, to the U.S. Mint at San Francisco, and it was this that triggered off the first rush in the spring of 1858.

Through no fault of the H.B.C., the amount of gold on the Fraser was grossly exaggerated, though the Company was to be blamed by many for their failure to strike it rich. In fact, San Francisco papers and shipping companies were the culprits, the original overstater being a trader called Ballo, eager to start an express company between the city and the Fraser.

The result was a genuine stampede, at least 25,000 converging on what was not yet the colony of British Columbia. Most of them came from California, others from Washington, Oregon and even Central and South America and Hawaii, the first ship reaching Victoria in late April.

Some travelled overland through Oregon and Washington and suffered accordingly from wild terrain and often equally wild Indians. Herman Reinhart was one of those who went this hardest way and confessed to his diary that he had never suffered so much in his life. There was another route through thick forests after landing in Bellingham Bay, Washington Territory, but this rugged trail, publicised by traders on the make, was abandoned when it was found that steamers that had served on the chief rivers of the Californian gold rush, the Sacramento and the San Joaquin, could reach Fort Yale on the Fraser. Much to their disappointment, the Bellingham traders had nothing to offer their patrons now except the chance to avoid a licence fee.

This could only be obtained in the early days of the rush at Victoria, where Douglas was proving well able to cope with the erupting situation. But before seeing him and his tough Chief-Justice, Matthew Begbie, at work, it should be noted that on the Fraser and Thompson, geography helped limit the size of the rush and made it easier to manage. Though gold was to be struck on sandbanks and bars of the

Miners at work near Sugar Loaf Hill, California, in 1852

Near Sugar Loaf Hill
Nevada County Cali
Early 50's

One of the earliest of all gold rush photographs, showing a miner working at his cradle in 1849

Fraser between Fort Hope and the junction with the Thompson, only further up both rivers were large amounts to be found. But supplies had to be taken up river and this, plus the hostility of local Indians who were themselves engaged in gold-seeking, did not encourage any but the most determined. There was little opportunity to use anything more ambitious than the rocker and the pan, the gold-bearing bars were small-scale and to the surprise of the Californians, the Fraser was too high. The first strikes occurred before it rose in the summer, and the snows kept the river high well into August, holding up work on the bars to such an extent that many left, not least to avoid what was clearly destined to be an icy winter. Added to this, rumours kept flying around of new strikes, which turned out to be non-existent, and morale, which was rarely high in 1858, slumped still further.

Meanwhile, Douglas had been having the most active year of an active life. An imposing-looking man, who believed in making his presence felt, he was officially only Governor of Vancouver Island, with no jurisdiction over the mainland. Showing the initiative which, with his other qualities, was later to be rewarded with a knighthood. He placed a licence fee of 10 shillings a month on would-be prospectors as a token of the Crown's right to any gold found, then, when the rush started, after an initial period of some discrimination against Americans, he levied fees on vessels entering the Fraser and added a tax on all goods, also concentrating on collecting the licence fee. This had now been raised to $5 and, inevitably, many managed to avoid payment. As to his anti-American measures, they were due to his concern for his Company's rights, also, perhaps, a natural irritation at the way Britain had been eased out of Oregon.

He lived in what Reinhart called a fine palace, and from it he strode out to confront troublemakers. One such, an American, objected to taking the oath of allegiance needed to get permission to hold land. 'Suppose we came and squatted?' said the American, not knowing his man.

'You would be turned off,' said Douglas firmly.

'But if several hundred came prepared to resist, what would you do?'

'We should cut them to mincemeat, Mr——; we should cut them to mincemeat,' was the reply.

It was important to establish a British way of doing things in the gold-fields, so Douglas visited them in May, 1858, not so much to throw his weight around–he allowed the miners at Hill's Bar to keep their 25-foot frontage claims–but to show them that he and his newly appointed revenue collector were in charge. He employed miners to build a trail to the Upper Fraser, and managed to placate the local Indians after miners had taken the law into their own hands when two Frenchmen had been killed. The miners had killed at least 30 Indians in two punitive expeditions, but Douglas, with a small detachment of soldiers and sailors, calmed the chiefs, forbade the whites to sell the Indians liquor, and selected two justices of the peace to see that there was no more trouble, then left with the tough Californians respecting him as a man and as a servant of the Crown.

Unlike some of his Australian counterparts, he was backed up by his Government at home, whose only objection was earlier discrimination against Americans in favour of the H.B.C. Otherwise, his reward was striking. The colony of British Columbia was established with himself as its first Governor. Naturally, he had to leave his Company post.

Reinhart was one of those who left the new colony rather than face the winter there. He passed through Victoria, where there were now several warships, and where, on his first visit to a church in eight years, he saw a British marine being buried with full military honours after being accidentally killed. There were some American miners present, well-lubricated and emboldened to scoff at the Redcoats lining the grave and firing their guns in salute. Reinhart and his more respectable fellow countrymen were deeply ashamed at their drunken jeers.

He noted the number of criminals, driven out of San Francisco by the vigilantes, who had turned up in Victoria, also some New York hard cases. They got nowhere in Victoria, but caused the authorities trouble up the Fraser. Fortunately, by now Chief-Justice Begbie was backed up by troops and sailors, not many, but enough to keep ruffianism on the gold-fields to a minimum. The chief crime was smuggling liquor.

Mining became more technical in 1859, and the licence fee was reduced in the miners' favour to £1 or £5 a year.

A home from home for the Forty-Niners

It seemed, though, that the British Columbian fields were fading fast until, in 1861 and 1862, new finds were made to the north on the Quesnel and in the Cariboo region. Thousands travelled to these wild, remote areas, while 165 Royal Engineers, as well as assisting Bebgie, blasted mountain sides to construct the most difficult sections of the celebrated Cariboo Road.

The biggest strikes were made at William's Creek, 400 miles beyond the head of navigation on the Fraser. 4,000 were working there by 1863, many of them engaged in tunnelling and shaft-sinking, and between 1862-64 some 200,000 ounces were produced from the Cariboo as a whole. The area was remote enough even when the great road was completed by the Engineers and private contractors in 1865, but before that transportation was something of a nightmare. Some carried their belongings on their backs, or on horses or mules, others used U.S. Army surplus camels bought by Frank Laumeister until their nasty habits–biting and kicking and their smell–so upset the other packers that lawsuits brought the practical scheme to an end. The trail got steadily worse, with paths often at an angle of 40 to 50 degrees. Fallen trees and morasses were among other obstacles until once again Douglas came to the rescue. It was he who gave orders for the 18-foot wide road to be built.

While construction was in progress Barnard's Express Stage Coach Line began operating and was later to run over one of the longest routes in North America. Less rapid, but essential to the smooth running of the mines, were the ox and mule trains that went up and down the Cariboo Road.

Men worked hard even by mining standards on the Cariboo. Though panning produced the first gold, the mines very swiftly became mechanised despite the colossal expense of getting machinery to them. Pumps had to be installed, which put costs up yet again. Best known of all the claims was the Cameron Claim where 'Cariboo' Cameron employed 80 miners at $10-16 a day. In 1863 it was yielding from 120 to 336 ounces a day, while in all it produced $1 million.

Victoria was again a boom town amid all these excitements in the north, though many of those who arrived there heard about conditions in the wilds and settled down in the city instead. Meanwhile, up in the Cariboo, a lively time was had by all off-duty, but a law-abiding atmosphere was the rule, as it was to be in the Klondike to an even more remarkable degree. Even the dancing girls, commonly called 'hurdy-gurdy girls' or 'hurdies', seem to have been a cut above the average, and winters were enlivened by entertainments in the log cabins, including debating, acting and singing. There was even a public reading room.

In the late '60s, the mining boom began to decline steadily, then, in 1871, came union with the Dominion of Canada on the understanding that a transcontinental railroad would be built. Gold went on being mined, and British Columbia remains an important mining area to this day. It was gold that converted a distant fur-trading area into an aspiring colony. But, meanwhile, many Californians had headed east instead of north to an area destined to astound the world.

Some 20 miles south of that mecca for quick divorce and gambling, Reno, lies a tourist attraction which has never quite become a ghost town, and which just a century ago had a population of 20,000. It is Virginia City, and in the 1860s and 1870s it was one of the most famous spots on earth.

For Virginia City read the Comstock Lode, whose metropolis it was. This fabulously rich vein of silver and gold was the biggest single bonanza of them all, and its historian, William Wright, who worked for the *Territorial Enterprise* in Virginia City, called his classic book about it, *The Big Bonanza–An Authentic account of the Discovery, History and Working of the World-Renowned Comstock Lode of Nevada.* On his paper he used the pseudonym 'Dan De Quille', also using it for his book, which first appeared in 1876. One of his colleagues on the *Enterprise* was the young Mark Twain, who also immortalised Virginia City in *Roughing It.*

These literary giants in the making were matched by mining giants, for such was the nature of the Big Bonanza. It was said of J. P. Jones, one of the bonanza kings, that he counted his dollars by millions, and that he had about five times as many millions as he had fingers and toes.

As early as 1849, gold was discovered in the area by a trader named Abner Blackburn in a ravine between Mount Davidson and the Carson River which was to become known as Gold Canyon. At the time this 'Washoe region', named for the Washoe Valley which was itself named for the Washoe Indians, was part of the Territory of Utah.

This and later finds were not enough to start a rush to such an inhospitable landscape, but in 1853, two brothers, Hosea and Allen Grosch, arrived from California, where they had had no luck, but where they had studied every aspect of mining from books. They struck silver ore in 1856, kept their secret while trying to find a rich partner, but both died in 1857. Hosea hit the hollow of his foot with his pick and died from blood-poisoning, while Allen, determined to return to California and earn some money for another expedition, died from exhaustion and exposure in the cruel Sierras. Their father was later to claim that they were the true discoverers of the Comstock, but without success.

Part of the Comstock was found in January, 1859 at the head of Gold Canyon, though no one realised it as yet, but the gold strike by four prospectors was a rich one, and the town of Gold Hill sprang up. The rush was a modest one still, but local laws were drawn up, including similar punishments to the ones meted out in California.

Enter Henry T. P. Comstock, liar and claim jumper, but the man whose name was to become immortal in the story of mining. It was now June, and this good-natured, accident-prone, unworthy man, who was to commit suicide in 1870, had been around the Washoe since the spring of 1858, when he took over Allen Grosch's cabin, how no one knows, as he told so many different versions of his story. On the day when two Irishmen struck it rich, Comstock managed to be around, as Dan de Quille memorably described:

> In the evening of the day on which the grand discovery was made by O'Riley and McLaughlin, H. T. P. Comstock made his appearance on the scene.
>
> 'Old Pancake' the liar's nickname, who was then looking after his Gold Hill mines, which were beginning to yield largely, had strolled northwards up the mountain, toward evening, in search of a mustang pony that he had out prospecting for a living among the hills. He had found his pony ... and came riding up just as the lucky miners were making the last clean-up of their rockers for the day.
>
> Comstock, who had a keen eye for all that was going on in the way of mining in any place he might visit, saw at a glance the unusual quantity of gold that was in sight.
>
> When the gold caught his eye, he was off the back of his pony in an instant. He was soon down in the thick of it all–'hefting' and running his fingers through the gold, and picking into and probing the mass of strange-looking 'stuff' exposed.
>
> Conceiving at once that a wonderful discovery of some kind had been made, 'Old Pancake' straightened himself up, as he arose from a critical examination of the black mass in the cut wherein he had observed the glittering spangles of gold, and coolly proceeded to inform the astonished miners that they were working on ground that belonged to him.
>
> He asserted that he had some time before taken up 160 acres of land at this point, for a ranch; also that he owned the water they were using in mining...
>
> Suspecting that they were working in a decomposed quartz vein, McLaughlin and O'Riley had written out and posted up a notice, calling for a claim of 300 feet for each and a third claim for the discovery, which claim they were entitled to under the mining laws.
>
> Having soon ascertained all this from the men before him, Comstock would have 'none of it'. He boisterously declared that they should not work there at all unless they would agree to locate himself and his friend Manny (Emmanuel) Penrod in the claim. In case he and Penrod were given an interest, there should be no further trouble about the ground.

The discoverers decided that they had better agree to save trouble.

> Comstock next demanded that one hundred feet of ground on the lead should be segregated and given to Penrod and himself for the right to the water they were using–he stoutly asserting that he owned not only the land, but also the water, and, as they had recognised his right to the land, they could not consistently ignore his claim to the water flowing upon it. In short, he talked so loudly and so much about his water-right that he at last got his hundred feet, segregated, as he demanded. This hundred feet afterwards became the Spanish or Mexican mine, and yielded millions of dollars.

Though de Quille did not mention it, the brazen Comstock also cut his friend, 'Old Virginny' Fennimore, in on the claim-jumping act, the latter having been one of those who made the strike at Gold Hill earlier in the year. Old Virginny did not own his new ground for long, Comstock buying him out for 40 dollars and an aged and blind horse. Later, when, as the Ophir mine, these precious feet became rich beyond belief, Old Virginny was to be heard lamenting that he had a $60,000 horse but couldn't afford a saddle!

Comstock and his partners worked away at their claims, producing about 30 to 60 ounces each a day, but finding much 'blue stuff' which in their ignorance they threw away, little realising that the Comstock was even richer in silver than gold. The exact sequence of events and who did what is uncertain, but the results are not in dispute. The Comstock's original owners, plus a newcomer, J. A. Osborne, sold out, thinking their mines would soon be exhausted. Given that they had no idea that they were potential millionaires, they secured what seemed good prices for the Ophir (O'Riley's and McLaughlin's first strike) and the Mexican mine beside it. This last went cheaply, Penrod getting $3,000 and Comstock–who had collected $11,000 for his share in the Ophir–taking two jackasses. It is calculated that in view of what the Mexican mine later produced, the animals were worth $1,500,000 apiece.

A brief farewell to Comstock is in order here. An account of his matrimonial misadventures can be found in *The Big Bonanza*. After his wife had run away with a strolling miner and he had lost everything in trading ventures, he wandered through the West hoping for another Comstock, only to end it all with his revolver near Bozeman City, Montana, in 1870. De Quille had the last word:

> The Montana papers said it was supposed that he committed the act while labouring under temporary aberration of mind, and this was doubtless the case, as his was by no means a sound or well-balanced brain.

Poor Old Virginny, also known as Old Virginia, and the alleged namer of Virginia City, also died violently. The alleged naming occurred when he was drunk, fell to the ground and broke his bottle. When he got up again he announced that he had baptised that ground Virginia. Finally, as de Quille wrote:

> He was killed in the town of Dayton, in July, 1861, by being thrown from a 'bucking' mustang that he was trying to ride while a good deal under the influence of liquor. He was pitched head-first upon the ground, suffered a fracture of the skull, and died in a few hours. At the time of his death he was possessed of about $3,000 in coin.

A Paiute Indian had the last word on both men, fortunately directing his sentiments towards the indefatigable de Quille:

> Hoss kill um Ole Birginey, Comstock he kill heself. Comstock owe me fifty-five dollar; Old Birginey owe me forty-five dollar! Me think ... maybe both time too much whisky.

The Paiute wronged at least one of his debtors. We have de Quille's word for it that Comstock 'was a man who drank but little.'

The 'blue stuff', dismissed as 'useless', was conclusively proved to be silver by two independent assayers about a month after the finding of the Ophir, and it was at once clear that a rich gold mine was even richer in silver. The first Washoe rush began, and Virginia City was born. Alas for the good story already narrated, the town was in fact given its name at a council meeting a full month before Old Virginny's alleged christening.

Some 4,000 arrived at the Nevada diggings that summer and autumn, many from California on hearing from the two assayers, Melville Atwood and J. J. Ott, how rich the finds were. They were respected men and miners, and no one doubted their stories. By the end of the season the Ophir Company had shipped out 38 tons of ore in boxes and sacks to San Francisco by mule train, and its value turned out to be $112,000.

Virginia City, meanwhile, spawned mud huts, canvas houses and cave dwellings, the homeless living in lodging houses of a sort that provided a bunk for a dollar a night and a spot on the floor for half that amount.

The winter was a rugged one, snow first falling in late November, and soon blocking not only the High Sierras passes, but even the road to Gold Hill. A few saloons and gambling houses kept up morale, but food ran short and not until March, 1860 did relief come, even though the first mule train was not the one most urgently needed. It brought liquor and bar fittings.

Now in ever-growing numbers the prospectors began to arrive, the majority of them on donkeys. There were soon 4,000 claims within a 30-mile radius of Virginia City, though few were worth possessing, and new techniques had to be learnt to mine the silver which made up more than half of the finds. Testing the quartz involved the addition of nitric acid, boiling it and adding salt, and other additions if there was any doubt. Not surprisingly, within a year 86 working companies had taken over the 4,000 claims. The Comstock, far more than California, was a 'big business' enterprise almost from the start.

And it was *big* business. Even before the supreme finds, the Big Bonanza of 1873, some $100 million of silver and gold were taken out from these West Nevada mines between 1859 and 1869, the Comstock lode alone producing 16\frac{1}{2}$ million in 1867. One of the first to make a fortune was George Hearst, who owned a sixth of the Ophir. His investment of a mere $450 was the foundation not only of of his success but of that of his notorious son, William Randolph Hearst.

Newcomers to the mines included Mexicans who had mined silver in the south, Chinese who tended to do everything that needed doing, except mining, and who were at least tolerated, and, later, German and Cornish miners. The independently-minded men swarmed hopefully over

A Wells Fargo chest. The famous firm transported the bulk of the gold found in the Western rushes

Nevada, while at the Comstock the skilled wage-earner predominated; dominating them were the 'bonanza kings'.

Before meeting some of them, the essential and fabulous facts about the Comstock may be summed up. After steadily producing some 16½ million dollars worth of gold and silver a year, by 1872 it seemed as if the great days were over. Then, in 1873, came the supreme prize, the Big Bonanza itself, 1,167 feet down, and most of it in the Consolidated Virginia and the California mines, a little being in the old Ophir. This greatest of all silver strikes was followed by installation of the latest equipment which in 1875 produced $38 million of silver. San Francisco, the terminus for this cornucopia, began to sprout millionaires, or so it must have seemed to its inhabitants. In all, between 1859-80, the Comstock produced $300 million in silver and gold. As for the site of the Big Bonanza, its richest ore was worth $1,000 to $10,000 per ton of ore, and in all the Consolidated Virginia and California made $150,000,000 and paid out $78,148,000 in dividends over 22 years.

The four chief mine owners were all Irishmen, John Mackay, James Fair, Jack O'Brien and James Flood. Mackay, the most remarkable of them and the least spoilt, lived until 1902, and when he died his business manager told the press: 'I don't suppose he knew within 20 millions *what* he was worth.' An old California hand, he had arrived at Virginia City in 1859 as penniless as his friend O'Brien. 'Let's enter like gentlemen,' said O'Brien, after throwing away their last half dollar for luck.

They worked as hard as any, were lucky, and proved born engineers and leaders. Even when Mackay was worth $60 million, he was regularly down one of his mines at 6 a.m. each morning. No one could dub these bonanza kings 'grocers'. The Big Bonanza enabled them to install the most up-to-date equipment in the world, not least because there had been a transcontinental railroad since 1869, and there were feeder lines as well.

Mackay, ex-Dubliner and ex-carpenter, never stopped studying rock, and could simply look at a sample of ore and tell its worth in silver almost as accurately as an assayer. Another 'king' was William Sharon, who built the biggest hotel in the world in San Francisco. As for Captain Sam Curtis, he appeared to be psychic, so remarkable were his finds, and volunteers flocked to work for him. James Fair, for all his $40 million, was often to be found in his miner's garb down below and, like his colleagues, would try anything new in the way of machinery. The once penniless O'Brien founded the Bank of Nevada.

Despite attacks by jealous rivals, the Bonanza Four were the finest and fairest group on the Comstock, concentrating on mining and deep development and never indulging in stock market manipulations, also feeling a real duty to their stockholders. As for the miners who worked for them

Orson Welles's filmic masterpiece, *Citizen Kane*, was inspired by newspaper tycoon, William Randolph Hearst, whose father left him a fortune based on gold mining, especially from South Dakota's Homestake mine

and the other bosses, their life was rugged if well-paid by the standards of the day. In 1867, their union enforced a rate of $4 per day. Seams which had started at the 4-foot level rapidly got lower until men were working at 1,500 feet and finally at 3,000 feet in some mines. Ventilation was therefore a major problem, gradually solved by machinery, while compressed air drills were installed from 1872. Such lighting as there was before electricity arrived after the Comstock's great years, was provided by candles and lamps.

Heat was the worst problem, with 130 degrees at 3,000 feet. In one mine it reached 150 degrees and the soaking miners had to work short shifts, then visit cooling stations to drink iced water and rub themselves down with towels dipped in the same water. Shifts were eight hours. Apart from several major disasters–a mine collapsing, an underground fire–accidents were not as common as might be expected. One regular danger was ascending from extreme heat to the surface, which sometimes caused fainting *en route* and falling out of the cage, while pneumonia was liable to strike the half-clad, profusely sweating miner.

Away from his work, there was plenty to occupy him–hardly surprising in view of the sheer size and vitality of the town. Some 10,000 reached Virginia City after that first bleak winter, C. H. Shinn noting:

> Irishmen with wheelbarrows; American, French and German miners with tools and heavy packs; Mexicans with *burros*; gamblers and confidence men on valuable thoroughbreds; Missourians struggling through the mud with their families and household goods in lumber waggons; drovers with hogs and cattle; organ grinders, Jew peddlers, 'professors' with divining rods and electric 'silver detectors'; women, even, dressed in men's clothing and usually under some gambler's protection. One saw youth and strength, illness and old age, cripples and hunchbacks–'all stark mad for silver'.

It followed that early days in Virginia City were on the lively side, though most authorities seem to agree that decent, law-abiding folk were in the majority through most of the town's history, decent in the Elizabethan sense, i.e., full of gusto and the joy of life within the law, and never dull. Of course, Mark Twain's version of events is more fun. His *Roughing It* is mostly concerned with life in Virginia City, give or take a little artistic exaggeration, as he knew it on de Quille's *Territorial Enterprise* on which he served as a reporter, and as he heard about it from earlier arrivals:

> The first twenty-six graves in the Virginia cemetery were occupied by *murdered* men. So everybody said, so everybody believed, and so they will always say and believe. The reason why there was so much slaughtering done was that in a new mining district the rough element predominates, and a person is not respected until he has 'killed his man.' That was the very expression used.

Not content with these high jinks, the first teacher packed a bowie knife and three pistols in class, threatening a pupil who whispered, then, when another threw a ball in the air during break, put a bullet through it. But good humour abounded on the Comstock, and the atmosphere was a generous one. Every sort of sport flourished, and miners who appear to have enjoyed Shakespearean plays enormously, supported blood sports known in Elizabethan London: bear fights and bear-dog fights. Humans fared better, a nice example of Comstock generosity being the way in which an elderly singer from Boston with a voice to match was encouraged to retire. Some of the patrons kept encoring her and hurling silver half-dollars at her feet, providing her with enough to prevent her ever appearing in public again.

Everyone read the newspapers, and when there was no news, reporters were expected to invent some. The great de Quille was a famous liar when space had to be filled. Once he described how the inventor of a cooling suit had gone to Death Valley to test it. He did not return, and friends found him frozen to death, having been unable to turn off the compressor on his outfit. On his nose was an 18-inch icicle, though the temperature was 117 degrees in the shade. *The* (London) *Times* picked up the story and suggested that these suits might be useful for the Army in India. Mark Twain, his colleague, pulled off some equally improbable hoaxes. He had come as Samuel L. Clemens to Nevada to act as secretary to his brother, the local secretary of state, but found that the post offered no salary. After failing as a miner, he turned to journalism, first as 'Josh', then adopting his immortal *nom-de-plume* of 'Mark Twain', which was based on the Mississippi steamboat phrase, 'By the mark twain' (two fathoms deep), for he had once been a riverboat pilot.

The rest of the world soon became aware that it was all happening at Virginia City, which by 1875 had a population of 20,000. When Charles Dilke visited the town in 1869–he is quoted at the head of this chapter–he noted that it had passed through its second period–'that of "vigilance committees" and "historic trees"–and is entering the third, the stage of churches and "city officers" or police.' He had clearly missed the good old days when the citizens expected 'a dead man for breakfast every morning.'

Getting supplies to Virginia City and the other mining towns across the Sierras from California was always a problem, even when toll roads had been built in the 1860s, before which donkeys had proved even more ideal than mules as freight carriers on the highest slopes, including Mount Davidson, which contained much of the Comstock. Camels were imported from Mongolia, but though the fifteen which survived the journey to Nevada were happy in the sands at the foot of Mount Davidson, the prevailing alkali burnt their noses, hurt their eyes and made their sores even sorer. Soon, these naturally bad-tempered creatures were behaving worse than army mules, not least because they were badly and harshly handled. They were left roaming around and were all dead by the late 1870s.

The main road from California to Virginia City was one of the busiest in the nation in the '60s, and stagecoach traffic on all roads was immense. Wells Fargo was the main express company and the leading transporters of treasure. Drivers had to be skilled to pilot stages down from the Sierras. According to Vice-President Schuyler Colfax, much more talent was needed than for being a Congressman.

None of Nevada's other mining centres remotely matched the Comstock lode, though some produced other minerals as well as gold: lead at Eureka, copper and lead at Cherry Creek. But by 1900, Virginia City's days of greatness were finally over. It had brought colossal wealth to individuals, to California and, especially, to San Francisco; it had helped finance the Civil War; it had produced so much silver that nervous European powers were forced to take steps to keep gold as their monetary standard; and it had enjoyed itself hugely in the process.

In striking contrast to the Comstock, which so rapidly became a highly professional mining area, Colorado's Rocky Mountains saw greenhorns stampeding hopefully, inflamed by travellers' tales and downright lies perpetrated in books and newspapers and by Chambers of Commerce in the Middle West.

These Rockies had few inhabitants in 1858. There were some thousands of Indians, including Cheyennes and Arapahoes, but apart from the Mormons at Salt Lake City and a few isolated military posts, there were scarcely any whites. Mountain men, seeking beaver when its fur was high fashion between 1810 and the early 1840s, had blazed trails through the Rockies, and some land-bound Argonauts had passed through them, though most had gone across the continent further north along the Oregon/California Trails. In 1850, a 120-strong party of whites, Cherokees and Negro servants had stumbled on a little gold near the future site of Denver, but they had pressed on to Eldorado in California.

In the summer of 1858, William Green Russell, a Georgian who had had no luck in California, and had then gone back home, returned with a group to the West and, in particular Colorado–then part of Kansas–to try for gold in the region of Pike's Peak. Others appeared and joined him, but most of them drifted away, leaving Russell and a dozen companions. These thirteen struck placer gold on the South Platte, then headed north, prospecting and hunting into Wyoming. And there were other strikes by other parties, enough to bring a few hundred to winter at Russell's original camp at Cherry Creek.

There was no justification for a major rush, but a combination of events triggered one off–the ones noted above. It is no libel to include local Chambers of Commerce among the culprits, for the Missouri River towns had endured a panic and a recession, and clearly a gold rush was the answer to their trading prayers. Russell was accused of trying to keep his gains to himself and his friends, when he reported cautiously and accurately about the strike, and, meanwhile, the papers were full of headlines like A NEW ELDORADO IN KANSAS. Towns hired agents to race through the area plugging it by word of mouth, and in totally false publications.

One such announced: 'Gold exists throughout all this region. It can be found anywhere ... In fact, there is no end

One of the first photographs taken in a mine by magnesium light. The miner is deep below the surface at Virginia City, Nevada of the precious metal. Nature herself would seem to have turned into a most successful alchemist in converting the very sands of the streams to gold.' And another gushed: 'At least as far as the summit of the Rocky Mountains, the journey is one of the most delightful and invigorating.'

Wilder and wilder got the claims, and it was no wonder that not only Kansans affected by the depression but thousands from elsewhere were seduced into stampeding, their rallying cry being: 'Pike's Peak or Bust!', which was not even the correct spot to aim for.

While those miners already in Colorado bitterly suggested that the originators of the false stories should be lynched, some 100,000 set out in the Spring of 1859, most of them knowing nothing of mining, few knowing exactly where they were going, and only a small number having been in California. It was a very different rush from the Comstock's.

They arrived to find a few towns laid out, as yet mainly on paper, notably Denver and Aurora on either side of Cherry Creek, but there was precious little else to occupy their time, and many started for home right away. Half their number were gone by the end of the year. And ominous rhymes were concocted:

Here lies the body of D. C. Oakes
Killed for aiding the Pike's Peak hoax.

Canvas tops which had borne the brave words, *Pike's Peak or Bust!*, now had a heavy line drawn through them and below was written 'Busted, by God!' and suitable slogans. As for Mr Oakes, he actually got away with nothing worse than a roughing up, but as his name rhymed with hoax, a fresh gave was dug and the rhyme written on it. He came across it and must have read the message with mixed emotions.

This was by far the greatest fiasco in North American mining history, but something was gained, for many would-be miners remained in Colorado to farm in its fertile high valleys and the plains to their east.

Needless to say, Colorado *was* rich in gold and other minerals. As early as January, 1859, George Jackson, who had been in California, found placer gold on a fork of Clear Creek, keeping quiet about it until April. Also in January, a Georgian named John S. Gregory struck gold on another fork of Clear Creek, the first important find in Colorado. Short of supplies, he fell in with a group led by Wilkes DeFrees from Indiana, who offered to grubstake him. This meant providing him with supplies in return for a stipulated share in any profits, and was to become a very widespread practice, the grubstakers usually being storekeepers. Unlike most of them, DeFrees went along with Gregory and on May 6 a really rich vein of quartz gold was struck at a spot which became known as Gregory Gulch. The footloose finder sold out for only $21,000, but it was he who started the real Colorado gold rush.

Soon, 5,000 men had reached Gregory Gulch, and in June came Villard and another journalist named Richardson, and, most notably Horace Greeley of the *New York Tribune*, who was to play the most decisive part in spreading the fame of the finds, for he was one of the best-known Americans of his day. Not only did he address a mass meeting of miners of the joys of temperance and right living, and the way to make Colorado a state, he and his fellow journalists gave maximum publicity to the finds in a joint report. The fact that they admitted it was not an easy spot to reach made no difference. Soon there were some hundred sluices at work near Gregory Gulch, and from all parts of the world a mainly youthful flood of hopefuls headed for the latest land of heart's desire. Only Californians held back, apparently, being more concerned at this time in British Columbia.

The eastern slopes of Colorado's Rockies turned out to be a 200-mile mineral belt, for other strikes were soon made, none of them fabulous, but big enough to keep up the excitement. However, only $25 million was extracted between 1858-67, which, considering the thousands that flocked in, was disappointing. And the years from 1864 to 1868 saw a slump. Colorado lost many of its immigrants and in 1870 had a total population of 39,000, only 5,000 more than in 1860, with Denver in the latter year having 4,759 inhabitants. It was fortunate that the area was to prove so suitable for agriculture.

Yet the Colorado mines, set in such striking landscapes, had the usual air of excitement about them and the usual problems. The criminal code formulated at Gregory Gulch was short and to the point, and included draconian measures for grand larceny: not less than 15 and more than 300 lashes on the bare back, a fine double the amount stolen, and banishment from the district.

Many were soon hiring themselves out to others for a regular wage, while finding somewhere to sleep was a major problem. Cabins were few and overcrowded and many slept in tents, wagons, or in the open beside camp fires in an area where nights were chilly even at midsummer. Vegetables were in short supply and therefore vastly expensive, so scurvy broke out. So did dysentery and a typhoid-like disease called mountain fever brought about, according to George Willison in *Here They Dug Gold*, by bad whisky and worse water.

Colorado was to become more a silver than a gold state, though. Samuel Bowles, who studied the area as closely as he did the Comstock, pointed out that 'there is silver in all the gold ores and gold in all the silver ores'. After

Denver, Central City was the most famous of the early mining towns and, like Georgetown and other centres, produced both. By the mid-60s many of the mines were in the hands of Eastern capitalists.

In terms of sheer wealth, Central City, centre of the area's gold-quartz mining, was the leading town, especially after swallowing two other early camps, Black Hawk and Nevadaville. Yet in 1860, it must have seemed as if California Gulch was destined for mining immortality, when Colorado's richest placer was found there. The leader of the lucky Georgians who struck it rich was Abe Lee, who shouted, or so the story goes: 'By God, I've got all of California in this here pan!' 10,000 were soon on the spot, but the diggings were almost finished by 1865 with no one realising that a fabulous silver strike was soon to occur.

By 1870, Denver was a junction for three railroads, and five years later it had a population of 135,000, a major city by any standards, though so recently a small and wild frontier town. Statehood for Colorado came in 1876, just a year before the spectacular silver strike at Leadville, not far from California Gulch, caused a world-wide sensation of almost Comstock proportions. In a book about gold rushes the temptation to linger in Leadville must be resisted, but it is fair to say that few mining boom towns got off to a wilder start, and soon Leadville had become the silver-mining capital of America, complete (by 1880) with 15,000 inhabitants, an opera house and 28 miles of streets. Lead, copper and zinc were also mined, and there were silver princes to match the Bonanza kings of the Comstock, the most notorious of whom was the semi-literate H. A. W. Tabor, whose mistress, the actress best remembered as 'Baby Doll', performed in the opera house he bought for her, while he, having made her the second Mrs Tabor, performed in the U.S. Senate. And all this, as John Hawgood observed in *The American West*, at 'over ten thousand feet above sea level!'

Actual opera came to the opera house in 1882, the year that the most exotic figure in mining history reached Leadville. It was none other than the young Oscar Wilde, who lectured on *The Ethics of Art*. The miners prefered the opera, though disappointed that in *Fra Diavolo* 'only two shots were fired, and only one man was killed.'

Wilde, happily, was elected the 'Prince of good fellows', not for his absurd choice of subject, but because he went down Tabor's Matchless Mine, then proved a better drinker than any of his new-found mining friends. It was a great occasion for Britain in general and Ireland in particular.

After Leadville came a major find at Cripple Creek, which had a gold strike in 1890. This new Eldorado put Colorado firmly back as one of the chief producers of precious metals. By 1894, gold was ahead of silver once again, while Cripple Creek had telephones and electric power and light.

The other major strike, a silver one, was at Creede in 1890, where the notorious 'Soapy' Smith, con-man extraordinary and town dictator, increased his growing fame. We shall meet him again and more fully at Skagway in Alaska, where he was the master villain of the Klondike Stampede.

From 1858 to 1922, Colorado produced gold worth \$666,470,261 and silver worth \$497,359,655. Only in one respect had its rushes after the original fiasco been unique. Because of the Rockies, transportation there had been more of a problem than in any other mining state. Perhaps British students of the West can claim one other unique aspect of the Colorado mines, the extraordinary number of Cornish miners there.

These skilled workers were known as Cousin Jacks, and were clannish, amusing, inclined to practical joking and a source of endless amusement to Americans, who yet liked them and the way they did their jobs. After a time they sent for their wives, who were christened Cousin Jennies. They were to appear in other mining areas, as they had appeared in Nevada, but Colorado was their chief home, and many of their descendants are there to this day. Some of the stories about them were mocking, others were self-mocking, but they all added to the flavour these tough men with their strange accent brought to the mines.

Sadly, many of the bawdiest stories have been lost because of Victorian reticence in print, but plenty have survived about these men who fascinated Americans, whom they called 'my son', 'my 'andsome', 'my beautay' and so on. One slightly racy one that has survived concerns 'tributing'. In some mines miners were allowed to lease a section of the workings, paying for it by extracting that much more ore for the company. Off duty, they could then work in their leased sections, and this scheme was called tributing.

It was naturally a popular scheme, and the Cousin Jacks were furious when one mine boss stopped tributing altogether. They collected in a saloon for a wake, and as one of them happened to be a lay preacher, they suggested that he might oblige with a prayer suitable for the occasion, one that might get their 'tributes' reinstated.

'Dear Lord,' asked the lay preacher, 'does Thee know Simon 'arris, superintendent of the Poor Man Mine? If thee know 'im, we wish for Thee to take 'im and put 'im in 'ell and there let the bugger frizzle and fry until 'e give us back we tributes. And when 'e do, dear Lord, we ask Thee to take 'im out of 'ell again, and grease 'im up a bit, and turn 'im loose, Amen.'

Gold Hill, Virginia City, Nevada, part of the fabulous area that produced the Big Bonanza

IMPERIAL
EMPIRE

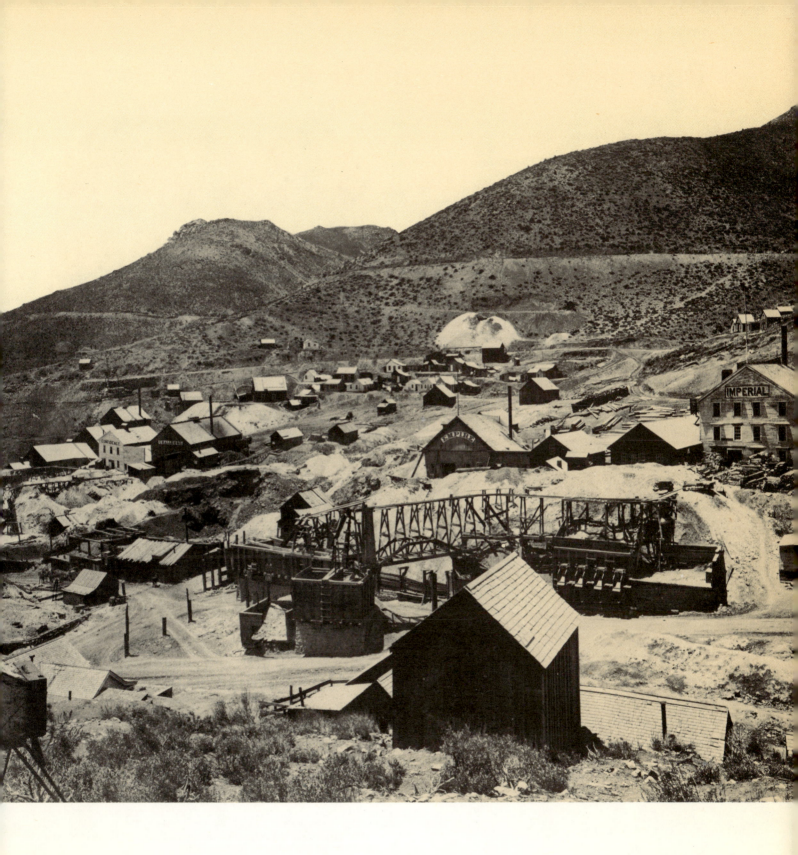

Rogues, Redskins and Rovers

What shall be done with these Indian dogs in our manger? They will not dig the gold or let others do it. . . .

Yankton *Press and Dakotaian,* Sep. 3, 1874

'The miners of Idaho were like quicksilver,' wrote the great American historian, Hubert Howe Bancroft. They 'ran after any atom of gold in their vicinity.' If mining camps in Idaho lived and died with extraordinary rapidity in the 1860s, the quicksilver quality was standard in most parts of the West. None of the bonanzas matched those of California, Nevada and Colorado, but gold was found in Montana, Idaho and the Black Hills of Dakota in considerable quantities, also in Arizona, Utah, New Mexico and Wyoming.

By late 1858, the Indian wars of the Northwest had finally ended, and the Columbia river, the main highway into the interior, had seen navigation developed by the military operations. Rumours that there was gold east of the Snake, the Columbia's main tributary, sent a miner named Pierce into the area, and he found colour near the south fork of the Clearwater in the summer of 1860. From then on events moved fast, not least because the Nez Perces, matchless Indians whose steady policy of friendship to the whites was to prove their undoing in the late '70s, allowed the mining centre of Lewiston to be built on their land.

Gold was found in the Salmon River Valley in 1861 and the Boise Basin in 1862, this last being the main mining area, producing more than all Colorado between 1863 and 1866. Thousands arrived from California, Oregon and British Columbia, and because the camps were so remote, they tended to get some of the wildest and woolliest characters. Those entrepreneurs who could face the journeys involved made fortunes out of goods, their wild women and wilder whisky, having spent a considerable amount *en route* paying other entrepeneurs who had erected roads and bridges and were collecting toll money for them. One such character had erected a 'bridge' across a stream by felling a tree and adding a few refinements. The result was guarded by an alarming looking character equipped with two six-shooters and a cheese knife, who announced to would-be crossers along the flattened top of the tree that the charge was $1 each for pack-animals and the same for passengers on foot. It was a simpler way of making money than panning.

By 1870, the boom was over for the placer miners who turned to other Meccas, leaving the claims to the less ambitious Chinese. Little machinery had reached these remote fields for quartz mining.

Idaho shares with Montana the dubious honour of having been the happy hunting grounds of one of the two supreme villains of the gold rushes (give or take some crooked financial wizards). As has been noted earlier, 'Soapy' Smith is being held back for his apotheosis in the Klondike Stampede, despite earlier feats in Colorado, but Henry Plummer, quiet, gentlemanly and of excellent appearance, belongs firmly to the Northwest of the early 1860s. He was one of the great con men of his age and his career tells us much about the period.

After emigrating to California from the East in 1852, he started respectably enough as a baker in Nevada City, becoming its marshal several years later. However, he killed a man who had dared to object to his wife being Plummer's mistress, and was imprisoned, but he was soon released, his friends having persuaded the Governor that he was a consumptive.

After various ventures in villainy, including murder and stagecoach robbery, he arrived at Lewiston in 1861, where his career took wings. Officially a gambler, he rapidly made himself the head of a gang of highwaymen–known as road agents in those parts–horse thieves and killers, who

This Minnesota schoolteacher never made it to California: the photograph was taken in his home town

swarmed over a huge area while he remained in town directing operations. Thomas Dimsdale, whose classic *The Vigilantes of Montana* contains much of what we know of Plummer, noted that it was 'only when excited by passion that his savage instincts got the better of him and that he appeared–in his true colours–a very demon.' He was also a notable seducer, but his cover held good in Lewiston, for when an irate citizen leapt up at a meeting after one of his friends had been robbed and murdered by the gang, Plummer was able to leap up after him, calm everyone down, and convince the townspeople that they must not take any hasty action.

He moved on to Florence after another killing, this time of a saloon-keeper who had denounced his inaction as cowardice, and to help increase his earnings, built two roadhouses called 'shebangs' in well-chosen spots frequented by miners and other travellers from the gold-fields. They were staffed with some of his choicest thugs.

Meanwhile, his agents in other townships noted any miners outfitting, then made out false bills of sale for their horses and saddles, which were rushed ahead of the miners to the shebangs. When the miners reached them, Plummer's men presented the bills and dispossessed the unfortunate travellers. Anyone arguing did not argue for long.

In the autumn of 1862, after some of his men had been recognised in a hold-up and, later, lynched, Plummer moved on to Bannack, Montana, his activities still unknown. He had to kill one Jack Cleveland who had fled with him, then fallen out with him over a woman, but he made it look like self-defence. In Bannack he reached the heights of his profession. He rapidly organised a 100-strong gang and made his headquarters in a ranch. Dimsdale had something to say about the spot:

> The headquarters of the marauders was Rattlesnake Ranch ... Two rods in front of this building was a sign post, at which they used to practise with their revolver. They were capital shots. Plummer was the quickest hand with his revolver of any man in the mountains. He could draw a pistol and discharge the five loads in three seconds. The post was riddled with holes, and was looked upon as quite a curiosity until it was cut down in the summer of 1863.

In May, 1863 he pulled off his supreme coup, getting himself elected as sheriff of Bannack. Now he had road agents and spies all over south-western Montana, including men in every saloon and store, who put secret marks on any stagecoaches worth robbing. And so as not to have his men wasting time on profitless hold-ups, he had them wear a special knot in their ties so as to be instantly recognisable to each other. They called themselves 'The Innocents'.

Naturally, three of his most desperate henchmen were elected his deputies, plus an honest man named Dillingham, a local citizen who unwittingly acted as window dressing. He had to be killed, but the killers were caught and two of them about to be executed when a combination of a faked letter to the mother of one of them, plus some skilled distortion of the voting, caused enough confusion for the culprits to escape on an Indian's horse. The third man had been acquitted. 'There is a monument to disappointed Justice,' said one of the guards, staring at the untenanted gallows.

Now came the high noon of Plummer's career. Killings and robberies increased as he enjoyed his secret status as a veritable Napoleon of crime. Extra recruits were obtained by enlisting disappointed miners, while Plummer himself, suitably masked, often went out on robberies with the boys. He would leave town as the Sheriff, disguise himself, hold up a stage, then return from 'official business' to Bannack. He was hoping to become a deputy U.S. marshal in due course, and tried to kill someone who refused to recommend him for the post.

The man who survived was N. P. Langford, one of the first to suspect Plummer's double life. Soon others were becoming suspicious also, then a young man named Tilden recognised Plummer while he was being relieved of his money. His friends advised silence, but the news circulated fast.

It never occurred to the arrogant Plummer that his reign of terror might be ending. On Thanksgiving Day he gave a splendid dinner in the town, including a turkey sent from Salt Lake City, yet by now many of his guests were well aware what sort of a man he was. It was a scene worthy of a man whose career has inspired many a Hollywood script writer.

It was Plummer's chief henchman, George Ives, whose brutal murder of a ranch hand named Thibalt brought about the downfall of the gang. The body was found and displayed in Nevada City, causing a sensation, and that very night 25 men met in the town behind closed doors and decided to take the law into their own hands. Ives and several others were captured, and the killer asked to be tried in Virginia City (Montana) to increase his chances of escaping with Plummer's help. No notice was taken of his request, and after a trial, he was hanged by the vigilantes.

Meanwhile, in Virginia City another vigilance committee was formed, and soon there were many more, and the miners began to breathe more easily.

At last, one of the Innocents, Red Yager, made a complete confession and explained how the Plummer gang operated. It did him no good. He was strung up and a notice was put around his neck reading: 'Red, Road Agent

and Messenger'.

Drunk with power, Plummer had not fully appreciated the danger he was in, and left his escape too late. On the night of January 10, 1864, with a bitter, icy gale blowing, vigilantes stormed his lodgings and captured him before he could reach for his gun. He and two of his deputies were dragged out, swearing in vain that they were innocent.

It was useless. The grim vigilantes hanged the deputies then, granting Plummer his death wish of a quick, efficient drop, dispatched the most notorious man in Montana. Many of his road agents suffered the same fate, others were banished and some fled. Vigilante justice continued spasmodically in Montana until the '80s, then faded away. Given that such methods were always controversial–though widely welcomed in Montana by miners and settlers alike–these vigilantes, organised in the first place in a truly desperate situation, were doing their territory, and later their state, good service.

Many Idaho miners had drifted to Montana before Plummer's infamous reign, and found there settlers who had arrived from the Mississippi Valley, from Colorado and Utah, many of whom joined the scramble for gold. The first strikes had been made as early as 1852, but the great finds were at Bannack (1862), Alder Gulch (1863) and Last Chance Gulch, later the state capital, Helena (1864). Virginia City, named after its more famous Nevada namesake, dated from 1863, being located at Alder Gulch, and had 10,000 people by 1865. By then it had settled down to respectability, the vigilantes having done their work well, and had the territory's first newspaper, the *Montana Post*, edited by English-born Thomas Dimsdale, whose book on the vigilantes has already been mentioned.

In June, 1865 James Knox Polk Miller reached Virginia City from Salt Lake City, as his vivid diary (University of Oklahoma Press, edited by Andrew Rolle, 1960) relates. He was startled to find that on Sunday every saloon, store and dance hall was in full blast, but was soon busy reading Dickens and Jane Eyre in his spare time (from book keeping), visiting the theatre, importing food from Salt Lake City as a side-line, and joining a new social club, a literary club and attending lectures. He sometimes resolved to give up billiards, drinking, smoking and riding, later merely limiting the amount of money he spent on them. He never staked a claim or panned for gold, but, like others of his day, was as much a part of the mining frontier as any prospector.

Montana's great placer mining years were from 1862 to 1876, and some $150 million in gold was produced, $390 million by 1950. Even more silver was produced by that last date, but this mineral-rich state's chief wealth has lain in copper, the most famous mines being at Butte.

The military and mining – Custer, leader of the Black Hills expedition, and the Cariboo Trail, built by Royal Engineers

Calamity Jane lies buried at Deadwood beside the man she wrongly claimed as her lover and/or husband, Wild Bill Hickok

The Indians have so far been shadowy figures in this story of the American gold rushes. We have seen the terrible fate of California's Indians and the defeat of the tribes of the Northwest who were 'in the way'. The Comstock had its Indian 'war' when the Paiutes were driven into the desert in 1860, and Colorado saw the infamous Sand Creek massacre by militia under Colonel John M. Chivington of Cheyennes and Arapahoes in 1864, to the delight of miners and settlers alike. Chivington, a minister by profession, only shocked local opinion when he later married his son's widow. By 1867, the Cheyennes and Arapahoes had finally been ousted from Colorado and by 1880, almost all the state's remaining Indians, the Utes, had been evicted.

The Nez Perces of Idaho and Oregon, good friends to the whites at all times, were also to find themselves in the way, which resulted in part of the tribe conducting an epic retreat under Chief Joseph of 1,500 miles in 1877. Time and again they defeated their pursuers until finally the remnants were cornered near the Canadian border which spelt safety. Even hardened Indian haters admire Joseph and his men, women and children, but the result was the same: removal to Indian Territory, where most of the Nez Perces died before at last the remnants were allowed to return to the Northwest, if not to their beloved homeland, so coveted by both miners and settlers.

But the greatest barrier was the Sioux. In 1865, the Government began to build the Bozeman Trail, named for John M. Bozeman, who pioneered a route from Bannack to Fort Laramie in 1863. This Government enterprise was deliberately started as a short cut to the Montana gold-fields. It caused Red Cloud's War, named for the great Sioux chief, as it cut straight through Indian land which was also their main hunting area, in what is now the heart of Wyoming.

The most famous incident in this war was the 'Fetterman Massacre' just outside Fort Phil Kearney in December, 1866. The fort had been built along the trail by Colonel Carrington's command, despite constant attacks by Sioux and Cheyennes, and Fetterman was a hot-headed, foolish young captain who knew nothing of Indians and despised them. He kept urging: 'Give me 80 men and I'll ride through the whole Sioux nation', and finally got his wish when a wood detail was attacked. His force was 81 strong.

Lured to his destruction by braves who included the young Crazy Horse, he and his men were wiped out. Though the fort survived, and though the Americans were to inflict two defeats on the Indians, who suddenly found themselves up against repeating rifles, this was a war that the Red Man won. In 1868, the Americans abandoned the Bozeman Trail and the whole of the Powder River Country 'for ever', and Red Cloud had the supreme satisfaction of seeing the troops march out of Fort Phil Kearney, which was then burnt to the ground by his warriors.

Long before these stirring events, gold had been found by the Sioux in the Black Hills of Dakota at an unknown date. Occasionally, they showed pieces at trading posts, but were warned by the celebrated missionary, Father DeSmet, to keep quiet about their finds or their land would be invaded. The traders kept quiet too, not wishing to have their pleasant life-style interrupted by a horde of prospectors.

But some prospectors had visited the hills–called 'Black' because of the look of their pine forests at a distance–at least as early as the 1850s, though the first known professional assessment was in 1853 by Lieutenant John Mullan, who led a detachment of men through them. An ex-Californian, he realised the possibilities so fully that he did not risk telling his men about them. Two Swedes found gold in 1865-66, returned to the hills and disappeared, after which various expeditions were organised, only to be turned back by the military.

These beautiful hills, some 120 miles north and south and from 40 to 60 miles broad, had earlier been dominated by the Cheyennes, who lost them to the Sioux, but were permitted to use them still. It should be stressed that Indians did not own land in the white sense. The earth was sacred, and how could one own one's mother? It was more in the nature of a sphere of influence, a hunting ground. And the hunting was good in the Black Hills. But by the mid-19th century they were also sacred to the Sioux, *Pa Sapa*, the sacred hills where Sioux gods lived. It was into this holy land that an expedition came led by Lieutenant Colonel George Armstrong Custer.

Born in 1839, Custer had proved a dashing cavalry commander in the Civil War, and later confirmed his reputation in the eyes of the public when he and his Seventh Cavalry surprised and destroyed Black Kettle's Cheyennes beside the Washita River one bitterly cold dawn in 1868. Ambitious, lucky, and a glory-hunter, he was not the Indian-hater of legend, even writing that if he were an Indian he would have been a 'wild' one. He and the Sioux thought alike in many ways.

The Black Hills expedition of 1874 consisted of ten companies of cavalry, two of infantry, three Gatling guns, 60 Indian scouts, a number of army engineers and some scientists, newspapermen, two miners and a photographer. The Press were in ecstasies, for they had been preaching invasion of the hills for years to improve and develop 'one of the richest and most fertile sections in America,' as the *Yankton Press and Dakotaian* put it just after the expedition. The article continued with an attack on the Indians, the opening of which is quoted at the head of this chapter. It continued in the same vein: 'They are too lazy and too much like mere animals to cultivate the fertile soil, mine the coal, develop the salt mines, bore the petroleum wells, or wash the gold.'

The last words are the significant ones, for settlers and miners alike had no doubt as to what the expedition was really about. Officially, it was a reconnaissance to obtain information about hostile trails through the hills and to explore them.

Custer found that the region was rich in pasture land and, more significantly, he found gold. Oddly enough, the official geologist was away climbing a mountain on the day when the two miners brought evidence of the first strike into camp, and was so incensed that he made no mention at all of gold in his final report. Yet within a few days the vast majority of the expedition were proving to their own satisfaction that there was plenty of gold to be had.

That was in early August, and by the 24th, the news first appeared in print in the *Chicago Inter-Ocean*. The entire nation soon knew that there was gold 'from the grass roots down', and soon the 'Thieves' Road', as the Sioux called Custer's trail, would be filled with miners and the countdown to Custer's Last Stand on the Little Bighorn would begin.

Legally, Custer had not been at fault, for a military reconnaissance would not have violated the 1868 treaty with the Sioux. But in the gold-crazed atmosphere, the expedition became a major event, and not merely because the always newsworthy Custer was leading it. Besides, it was he who invited miners to go with him, who sent out the first report of gold by the scout, Charley Reynolds, and who later announced that the Black Hills should be commandeered.

The first group of prospectors under John Gordon reached the heart of the hills just before Christmas, having evaded the army and the Sioux. They were not evicted until the following April–they took some finding in snows and temperatures 40 degrees below zero, and the first force to try failed to do so. The Gordon party had by now sent out news of their finds and, despite a strong statement from the Secretary of War, many groups were formed in every part of the States.

The army's task was hopeless. A patrol under Captain Edwin Pollock tried to keep men out of the western side of the hills, but those they caught and handed over to the civilian authorities were at once released. One man was thrown out of the hills four times, while many more headed in from the east. Nearly 1,000 illegal prospectors were hard at work by July, 1875. The Government by now had sent in another scientist escorted by 100 men, partly because of the negative report the first scientist had sent in out of pique. The result was a guarded but positive statement that there was certainly gold but not on Californian or

Australian standards. However, rumour had it that the scientist, Walter Jenney, had found enough placer gold in one creek to wipe out the national debt.

All this while, the Government was trying to buy the Black Hills from the Sioux, without success, while the soldiers kept up their removal tactics in the hills. To reassure the Indians, this policy was stepped up by a force under General Crook, who within a decade was to be the finest Indian-fighting general in the West, and also a good friend of the Indians. On this occasion, he showed that his diplomatic genius extended to miners as well, for he got many of them to clear out after a meeting where it was agreed that all claims held would be guaranteed for 40 days after the Black Hills were finally made over to the Americans.

In just a little over a year, the Sioux, though they had inflicted a shattering defeat on their enemies, had finally lost their beloved Black Hills for ever. Before concentrating on the miners who gained the great prize, the political and military facts must be outlined. Negotiations started and collapsed and the Sioux themselves were divided, until finally, in the late autumn of 1875 a directive went out to all Indians to return to their agencies by January 31, 1876, or be regarded as hostiles. Many never got the message, for the Sioux scattered widely each winter, many more had no intention of coming in. So in March the campaign against them began, reaching its climax on June 26, when Custer and his Seventh Cavalry, part of General Terry's Dakota Column, which itself was part of a three-pronged attack on the Indians, stumbled upon several thousands of Sioux, Cheyennes and Arapahoes under Sitting Bull, Crazy Horse, Gall, the Cheyenne Two Moon, and other chiefs.

Custer's legendary luck had run out that blazing hot June day, for he had fatally divided his command into four detachments, then he and 200 men had galloped ahead into their valley of death and been wiped out in a bitter fight that lasted about an hour. The rest of the regiment under Major Reno and Captain Benteen fought their own battle seperately, little knowing that total annihilation had overtaken Custer and his men.

It was the last stand of the Plains Indians as well, for never again could they gather in such strength. Crook and others conducted mopping-up operations, while Sitting Bull and many Indians escaped to Canada. Finally, Crazy Horse gave himself up, later to be murdered 'trying to escape'. The Sioux, those who had survived the campaign and those who had stayed on the reservations, lapsed into sullen misery. The final death throes of a proud nation were to come later, at Wounded Knee in 1890.

Three months after the Little Bighorn, the Black Hills had finally been signed away by many Sioux chiefs. Legally, it was not foolproof, but the facts spoke for

Sitting Bull of the Sioux, whose sacred Black Hills were invaded by Custer's expedition

themselves. The Indians could no longer hold them. Even Crook's 1875 triumph had not cleared the hills, for many miners had stayed behind, some simply because they knew nothing of his meeting with the rest of them. By the end of the year gold was being mined in Deadwood Gulch, a thickly forested area of the hills, which was soon to be world famous, even though the claims were by no means sensational or extensive. Many men extended their placer mining to other gulches, but some men struck it rich beside the booming little town.

It attracted some notable visitors, the most famous being Wild Bill Hickok, the 'Prince of Pistoleers'. Much abused by the revisionist school of modern historians, and alleged to have killed more than 100 men, he seems, given his violent times, to have been almost the equal of his legend, while his best biographer, Joseph G. Rosa, has found definite proof of only seven killings.

He came with a bunch of prospectors, hoping to make a strike big enough to let him retire. Lately married–*not* to the notorious liar and Frontier character, Calamity Jane–he never saw his killer, a nonentity who crept up behind him on August 2 1876 in a saloon, while he was playing cards, and shot him dead. The killer, Jack McCall, was later hanged, his motive for the cowardly shooting going with him to the grave.

The first mining town in the Black Hills was Custer, where the 1874 expedition had dug for gold *en masse* and where Crook had held his meeting. It was sited near French Creek, where the two miners on Custer's expedition had struck gold, and by March, 1876 it had enough citizens to organise the 125 strong Custer Minute Men, whose job in an emergency was to drive off Indians. The town soon lost its supremacy to Deadwood, but remained a useful halt for those going into the hills.

Deadwood was laid out on April 26, 1876, getting its name from the number of dead trees in the area at that time, caused, as was later found, by prehistoric pine beetles. The original single street was sited in a dead-end canyon and wound its way past tree stumps and potholes, and was thick with dry dust or mud. Ox-drawn wagons caused traffic congestion for hours. It was one of the wildest of all mining towns in its early days, a suitable setting for its famous, though fictional, superman, Deadwood Dick, the delight of the young and young in heart, and the despair of Sunday School teachers. The first story by Edward Wheeler came out in 1877, geographically vague, as if anyone cared. Flesh and blood visitors, apart from Wild Bill and Calamity, included California Joe, a notable scout and a friend of Hickok, Wyatt Earp, and, in 1882, Doc Holliday.

Apparently, the gunfighting dentist saved a miner's life in a saloon, when the bar-tender took exception to the little man's refusal to have another one. Finally, the irate bar-tender opened a drawer and went for his gun. Seeing that he meant business, Holliday drew his .45 and deliberately shot the man through the wrist.

His good deed brought a crowd, as his victim was hollerin' more than somewhat, and it seemed as if Doc might have some explaining to do. He did it simply, folding his arms and saying: 'Gentleman, I am Doc Holliday of Tombstone.'

Back in the '70s, in even wilder times, the theatre was one of the most respectable–and popular–entertainments. The main playhouse presented drama and comedy given by Mr and Mrs John S. Langrishe, while at the Gem a wide variety of shows were presented, including *The Mikado* and–for the first time–Sioux Indians performing dances for the palefaces. After the show, the (white) 'actresses' turned dancing partners, drinking partners, and, for those members of the all-male audience who wished it, sleeping partners as well. Up to $10,000 a night were taken in the good years.

There was endless gambling, a red light district, a school in 1876, followed by others, and an outdoor preacher on a dry-goods box, who was doing well until killed by Indians. And, wonder of wonders, there was a Chinese community which was actually spared racialist prejudice.

Four miles from Deadwood, now a major tourist attraction, the famous Homestake mine remains as active as ever, making South Dakota the U.S.A.'s leading gold-producing state to this day. It is situated at Lead (pronounced Leed), where placer gold was first struck in February, 1876, by Thomas Carey. But the great find was made by Moses Manuel on April 9. He was one of four partners, another of whom was his brother Frederick, but he was alone when he made the strike, he and his partners giving it its name. The four soon found a 200 lb lump of quartz, the most valuable ever found at the Homestake, and continued working, only to sell out to the representatives of George Hearst in 1877 for a disputed amount, either $70,000 or $105,000. As we have seen, Hearst was already rich from his Nevada holdings, but it was the Homestake that made his fortunes soar. Comstock methods were used after open cuts gave way to shafts and tunnels, and the Homestake's techniques kept on improving down the decades. By 1900, the various mines of the Homestake Mining Company had extracted over $59,000,000, by 1931, over $233,000,000, by which time shafts were more than 2,200 feet deep and there were more than 60 miles of tunnels running from them.

All this was very professional, a far cry from '76, when the sound of pick and shovel and rocker was heard in the Black Hills. One George Stokes wrote to the Denver

Tribune about those early miners:

> Where can the miner of olden times have gone? . . . The miner of today is a bitter disappointment to Bret Harte's readers. Common matter-of-fact fellows; I grieve to say that even the buckskin shirts, long hair and bowie-knives are falling into disuse. One sees no eight-foot neck chains nestling against coarse woollen shirts, no three-ounce nugget breastpins, no betting of oyster cans of gold on the sex of a horse three blocks away. It is true that the miner of today loves whisky, cards and women; but as compared with the Forty-Niner of California or the Fifty-Niner of Colorado, he is a hollow mockery, and a fraud. He is close and calculating, refuses to be swindled, and as a rule expects his money's worth when out for a lark.

Maybe! The local villains were certainly not frauds in the '70s.

They were helped by the Black Hills themselves, which gave ideal cover to road agents, for not until the early '90s did the railroad reach Deadwood, though local narrow gauge lines were built to haul ore and timber. In the early days, with the railroad some 200 miles away, the stagecoach was all important. The stagecoach saga is a book in itself, and giants like John Butterfield, the great Ben Holladay, and the famous pair, Henry Wells and William Fargo, who took over from him when the Napoleon of the West switched to railroads, and who carried more treasure than any other Western line, cannot here be given their due. But no account of the Black Hills rushes can fail to salute the Cheyenne and Black Hills Stage company and its legendary Deadwood stage.

As early as January, 1876 there was a stagecoach service to the Black Hills, taken over in February by Messrs Gilmer, Salisbury and Patrick, who made Luke Voorhees their superintendent. He was a very experienced miner, who had actually been the first to strike gold on British Columbia's Kootenay River in 1864, and who was also an experienced cattleman. He was in charge of 30 Concord coaches and 600 horses, and the passengers that first winter wore buffalo robes to keep themselves warm. In April a

Deadwood, Dakota Territory, in 1876, when it was a booming mining town with less than adequate roads

direct 178.5 mile route from Cheyenne to Custer was opened and in September the line was extended to Deadwood. Other lines headed to the magic spot from settlements in Dakota, but it was the daily stage from Cheyenne which never failed to cause excitement, especially when Johnny Slaughter, cracking his whip and yelling, headed down the street at full speed, the coach swaying and the team sweating. Suddenly, he brought his coach and six white horses to a screeching, slithering halt while the onlookers cheered the splendid show.

The journey took around 50 hours in dry weather, from 60 to 70 or more when the route was muddy. Meanwhile, stories of the wealth in the Black Hills became an open invitation of highwaymen to move in. Agnes Spring Wright in her classic and entertaining *The Cheyenne and Black Hills Stage and Express Routes* shows how press reports fanned the flames of greed. One Billy Gay, a rich placer miner had a watch chain some two feet long, composed entirely of gold nuggets 'from the size of a pea up to the size of a hickory nut, with two or three larger ones for pendants'. And the papers reported the size of treasure shipments. Before veritable war wagons were built, gold dust was carried on ordinary runs in a steel container, heavily locked and fixed to the floor of the stage in some way.

Though the Sioux were undoubtedly responsible for some of the crimes in the area–horse thieving, cattle rustling–they were being blamed for white men's crimes. This became clear when two mail carriers were attacked, killed and scalped, but only registered mail was taken from their sacks. And only very smart Indians could have opened the gates of a corral beside a stage station with a duplicate key!

That Spring one of those who rode shotgun on a treasure coach was Wyatt Earp, who had come to Deadwood from Dodge to find out the prospects for mining and gambling, but had freighted fuel instead. His version, 'as told to' his doting, imaginative biographer, Stuart Lake, was dramatic enough, for all that nothing of note happened to stop the gold getting through. That Spring 'cleanup', the moving of treasure when the winter was over, saw $200,000 worth of gold shipped safely to Cheyenne.

But on March 26, Voorhees received a telegram from Deadwood reporting the murder of Johnny Slaughter, though the 'five masked men who did the deed' failed to get the loot. The murdered man was given a grand funeral in Cheyenne, where his father was Chief Marshal J. N. Slaughter, not to be confused with the famous cattleman and sheriff of Tombstone, John J. Slaughter.

In 1878, the Cheyenne and Black Hills company, rightly assuming that after a quiet winter the open season for gold robbers would soon start again, had two special armoured coaches built, the first into action being 'The Iron Clad' or 'Monitor'. Later 'The Johnny Slaughter' took the road. Steel plates almost an inch thick lined the interiors, and could stop any rifle bullet fired from 50 feet. The steel chests were bolted to the floor, and, when outbound from the Black Hills, no passengers were carried, just four guards. Two more were on horseback ahead, and two behind, arms consisting of rifles, sawn-off shotguns and six-shooters.

Ordinary robberies continued, then, on September 28, 1878, 'The Monitor' was robbed some 20 miles south of Deadwood at the Canyon Springs stage station, though the hold-up is often called the Cold Springs robbery after a larger station nearby. Five bandits reached the station ahead of the coach, captured it, and bound and gagged the staff. In mid-afternoon the southbound stage arrived and pulled up where a stock tender named William Miner should have been ready to help change the horses. He had been thrust into a grain room. Gale Hill, riding beside driver Gene Barnett on the boot, called to the missing man, then jumped down, and moments later a hail of fire hit the coach, killing one of the guards and also a passenger, who, against company regulations, had been allowed to come on the run.

Pressure from miners and settlers forced Chief Joseph and his Nez Perces to beat an epic fighting retreat from their homeland to the Canadian border

The mounted guards reached some trees and kept up a heavy fire, but the bandits had captured the driver and, making him into a shield, shouted that he would be killed if his friends did not cease fire at once. The guards agreed and retreated, and the road agents went to work on the safe, carrying away $27,000, mainly in gold bullion, after several hours' work to open it, the station hands having been tied to trees.

Voorhees offered a $2,500 reward, and plenty of citizens in Wyoming and Dakota left the mines and other employment to ride the bandits down, few having the remotest idea where to look. But after six weeks, three-fifths of the money was recovered, and in a general purge of road agents, many were imprisoned and some were lynched. The four agents who had survived the robbery, one having been mortally wounded, were probably all accounted for.

Less glamorous than the stage was the freight business, run as we have seen it in other areas. But the Black Hills had one unique attraction, a lady bullwhacker named Madame Canutson, an amazing women who not only drove her ten yoke of oxen regularly from Pierre to Deadwood and back, but literally carted her one-year-old infant around while Mr Canutson looked after the old homestead. No one disputed her skill in dealing with her oxen, who were on the receiving end of language ripe enough to round off her qualifications as an early exponent of women's lib. The nearest equivalent to Deadwood was Tombstone in Arizona, but that was a silver strike. Gold was found in Arizona, most notably at Wickenburg in the 1860s, the name coming from Henry Wickenburg, who discovered what came to be called the Vulture Mine, but sold out for $85,000 a property that was to produce $3 million. Both Arizona and New Mexico have their share of Lost Mines, where men not so unlike the men of the Comstock, of Last Chance Gulch, Leadville and the other fabled spots, still search for gold. They die for it, too, sometimes by human hands. The Lost Dutchman Mine in the grim Superstition Mountains of Arizona has lured men to their deaths nearly every year of this century, and many of those deaths have been murders and some beheadings. Such is the lure of gold.

But these are not rushes. There is one more North American rush to be related, not the greatest of them, but certainly the grandest because it pitted men against nature as not even Colorado's had done. But before taking the last great trail–to the Klondike–there is one more country to be visited, South Africa.

The railroad and the Wells Fargo Express Station, potent symbols of the mining towns of a century ago

Riches on the Rand

Splendid reefs have been discovered on the High Veldt between Klip River and the Witwatersrand.
The Pretoria *Volkstem*, June 28, 1886

He was South Africa's only Forty-Niner, or so the modern historian of the South African gold rush, Eric Rosenthal, boldly claims for Pieter Jacob Marais. He was 21 when James Marshall made history, and heard about the Californian Eldorado when the news reached Cape Colony some ten months later. He sailed away on April 2, 1849, reached Liverpool in 53 days, and finally landed at San Francisco in January, 1850. By December, 1851, he was Australia bound with $376 and a bowel complaint, and in April, 1853 was back in Cape Colony, very experienced but not having made a fortune.

He was not to make one in South Africa either, though he deserved to, for he typified the restless breed who have dominated this book. He was not even the first to find gold in South Africa, for it was definitely found–earlier rumours apart–in 1803, by a Dr Heinrich Lichtenstein. But it was Marais who found gold on the fabulous Rand after crossing into the Transvaal, then the South African Republic, in September, 1853.

This was 100,000 square miles of frontier country with the biggest white settlement, Potchefstroom, having only some 600 inhabitants, and it was in the Transvaal, on the northern slopes of the Witwatersrand, that he made his modest but historic finds. 'Found a few specs [sic] of gold in the River Crocodile,' he noted in his diary, on October 7.

Other small finds followed, which resulted in the Boer leaders asking him to meet their Parliament. These conservative farmers were confused by the finds, fearing a flood of foreign (i.e. non-Boer) diggers, but wanting the prosperity that gold might bring, and they compromised by offering Marais a reward if he struck it rich, but insisted that he should not leak news of such a find to foreigners. In the event, and after he had been lucky to survive a tribal rising, he retired to the Cape, dying in 1865 without worrying the suspicious Boer rulers.

But news of his adventures had inevitably leaked out, while in Natal other prospectors were searching, spurred on by rewards being offered, and also by news of the Australian finds. Yet by 1867 nothing sensational had occurred when the situation was transformed by the discovery of diamonds.

Ironically, this discovery was to set off a rush far more typical of a gold rush than the actual one which was to occur in 1886, and which, as we shall see, was to be scientific and technological almost from the start.

The diamond fields, which since 1868 have yielded around £1,000 million worth of gems weighing some 100 tons, were first revealed by a find in the Hopetown district of Cape Colony, and soon stones were being found elsewhere, prospectors in 1869 concentrating along the banks of the lower Vaal. That year, Australian diggers, arriving from Natal, began the novel practice of digging for diamonds, shovelling the gravel into cradles, picking out the coarser stones by hand and, using Californian and Australian techniques, hoping to find gold as well as diamonds. They had little luck, but then struck it rich in a gravel hummock. Gardner Williams, in his standard *The Diamond Mines of South Africa*, wrote that 'such a discovery could not long be concealed from visiting traders and roaming prospectors, and before three months had passed some prying eye saw half a tumblerful of the white sparkling crystals in their camp.' A rush to Klipdrift began, made all the more feverish because a combination of shortage of money, drought, war between the Orange Free State and the Basutos, and an end to railway building, had brought about a depression. Diamonds, then gold, were to put an end to even the possibility of depressions, for all the problems that came with them–problems with which South Africa still has to contend.

By late 1870, there were some 10,000 diggers on the Vaal, the majority white and from Cape Colony. Ironically, only two years before the great geologist Sir Roderick Murchison, on being shown the very first diamond found in South Africa, told John Blades Currey, later a mining magnate: 'My Dear Sir, If you tell me this diamond was picked up at the Cape of Good Hope, I am bound to believe you, and I do, but if you ask me to say, on the strength of such an isolated fact, that the Cape is a diamond-producing country, I must decline. Indeed I will go further: I will stake my professional reputation on it that you have not got the matrix of the diamond in South Africa.'

Even the briefest account of the diamond rush, which was to prove Sir Roderick so wrong, must note certain features of it, for the gold rush that followed cannot be understood without it. South Africa was suddenly pitchforked into an industrial revolution, with Whites flocking into the diamond fields from overseas as well as southern Africa, while thousands of Africans, poverty-stricken, displaced and unsettled by long years of intermittent wars, flooded in as well, hoping for money and guns.

The racial situation which emerged can hardly be claimed as unique, though it set a pattern that has lasted to this day, transferring the social attitudes of white South Africans to an urban situation. In the early days of the rush, heavy work at some diggings was entirely done by Whites, but some diggers arrived with their Black servants to do the manual labour, while others found Africans on the spot. News of wages of 30 shillings a month plus food reached even distant kraals, and Gardner Williams vividly described what happened next:

First came the neighbouring Griquas, Koranas and Batlapins, with Basutos from their southern reservation, followed by a stream of Zulus, Mashowas, Makalakas and Hottentots, and Kafirs of a hundred tribes, ranging east to the Indian Ocean and far north-west into Namaqua and Bechuana lands and north-east into Matabeleland and the regions beyond the Limpopo and the Zambesi. The white diamond-seekers were willing to pay, for a few months' hunting for little white pebbles, enough to buy a cheap gun and a bag of powder and balls–most precious of all earthly things in the eyes of a roving African. Then the white camps were lively, humming social resorts, abounding with good food and tempting drink, where black men were welcome and well protected. So the natives swarmed in faster and faster.... Some of this swarm could be persuaded to remain at the mines for a year or more and work quite steadily; but most drifted away at the end of a few months, or as soon as they were able to get their coveted guns and powder pouches.

It was only a matter of time before Blacks were the workers and Whites the overseers, for Africans stood little chance when faced by discrimination and by their own lack of technical skills. The colour they brought to the camps was indeed unique. 'No mining camp on earth ever before held such a motley swarm of every dusky shade, in antelope skins and leopard skins and jackal skins and bare skins,' wrote Gardner Williams, who noted how with or without the aid of rum 'they might dash off at any moment into some wildly whirling reel or savage dance.' When Britain annexed the diamond fields in 1871 Africans and Coloureds were no longer debarred from holding digging licences, but few seem to have done so.

The heart of the diamond rush was Kimberley, named for the Secretary of State for the Colonies, where in 1869 a stone worth £25,000 had been found at a spot then known as Colesberg kopje. Here a turbulent frontier boom town exploded into life to work the mines which were to become the supreme source of diamonds in the world. To it came expert diamond buyers, and lesser figures in the trade, one of whom was to become world famous, the Whitechapel-born Jew, Barnett Isaacs, who rose to power as Barney Barnato. The even more remarkable Cecil Rhodes, the parson's son who was to prove a swashbuckling politician and a financial genius, arrived in Kimberley in 1872 at the beginning of a 30-year career rarely matched in the history of British Imperialism. In a single decade he welded some 70 companies into the all-powerful de Beers organisation, and within another decade had acquired the Premier Diamond Mine, and joined with Alfred Beit and Barnato to become the De Beers Consolidated Mines Ltd. The age of the individual digger was long past and in its place was a benevolent autocracy, controlling South Africa's output of diamonds and their world price. Diamond-stealing, a problem from the beginning in the industry, had been cut to a minimum; the Blacks, for all that they were second-class citizens, were by no means repressed, though, incidentally, forbidden to drink or gamble. This was big business controlled by a Big Brother.

But it needed gold to make the South African economic success story complete. Diamonds had made Kimberley a capitalists' paradise and created a system in what had recently been a wilderness that tycoons in London, New York and elsewhere could recognise and respond to. Meanwhile, the search for gold in significant quantities had been continuing down the years since the '50s, gathering momentum in the early '80s with considerable finds–and rushes–especially at Barberton on the borders of Swaziland. Yet the word 'considerable' is only used to compare the finds with past ones, for on all the evidence, South Africa was not to be compared with America or Australia as a land of gold. It was reasonable to keep on looking, for money had been made, but no one could imagine that the searchers had been crossing and recrossing an ultimate

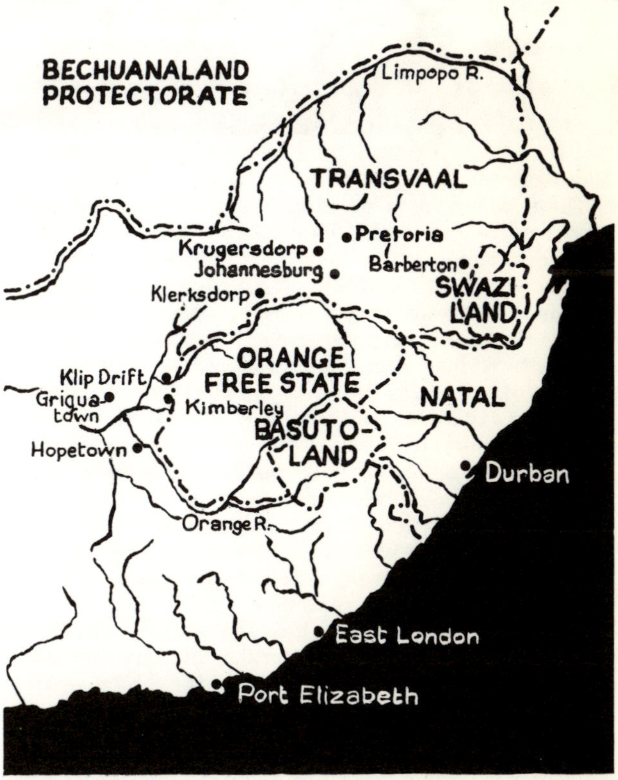

Gold Rush, South Africa

Eldorado.

There was and is nothing like the Rand gold-fields anywhere in the world. They extend some 170 miles by 100 miles, which make them the greatest in extent of all gold-bearing areas, and though the early mines have now been mainly worked out, newer ones appear as inexhaustible as the first finds once seemed. Well over 800 million ounces have been mined from this richest spot on earth where gold has been found in extraordinary uniformity, and yet the average gold content per ton is the lowest of all the earth's gold-producing areas, for the gold is microscopically fine. Therefore the gold rush to the Witwatersrand was no place for individuals almost from the start. It was less of a gold rush than any in this book, less of one, indeed, than the great diamond rush to Kimberley, for the quickest stampeders might be said to be the capitalists, headed by Cecil Rhodes. At hand was all the capital needed to develop the finds, the railways, the heavy machinery, the skilled workers and, from Natal, cheap coal, while African workers came in their hundreds of thousands, lured by the N.R.C. (Native Recruiting Corporation Ltd.) which inscribed magic words above one Zululand depot: 'Lovers of money, lovers of cattle, the road is easy to go to the City of Gold; here is the office.'

The City of Gold was Johannesburg, though just who Johannes was remains a matter of controversy. It was in the Transvaal, which had been annexed by Britain in 1877 at a time of public bankruptcy and danger from the Zulus. The Boers, resenting their loss of independence, had declared themselves free, had defeated the British at Majuba Hill in 1881 and had become self-governing, but under the suzerainty of the British Crown, in 1884.

Three Georges, Walker, Harrison and Honeyball, the first two born in England, the third half-British and half-Boer, stand out in a story that has no Marshall or Hargraves, for if there was a historic moment in an area where no single individual ranks as the discoverer of gold, it is usually regarded as the Sunday morning of February 7, 1886.

The three Georges were building a cottage for a prospector named Fred Struben. On the Sunday, Walker brought some banket (local auriferous conglomerate) to the homestead of Honeyball's aunt, Mrs Oosthuizen, where Honeyball was living. 'He borrowed my aunt's frying pan in the kitchen.' Honeyball later related, 'crushed the conglomerate to a coarse powder on an old ploughshare,

Exit the prospector with his pan. Enter machinery underground in South Africa and hydraulic mining in the American West

Artist's impression of the early days of the gold strikes in the Transvaal in 1887

and went to a nearby spruit [rivulet] where he panned the stuff. It showed a clear streak of gold.'

The next morning Honeyball found an outcrop of conglomerate, half a mile from Walker's spot, and on his Aunt Nellie's land. He showed it to Fred Struben, who annoyed him by telling him it was valueless, then came upon one Godfray Lys, who was more pleasant even before he saw what Honeyball was holding, when he became certain that George had made a great find. That night, Honeyball returned to his aunt's home, which was at Langlaagte. He told his Aunt Nellie not to sell her farm cheaply, as he and George Walker and Godfray Lys had reason to believe that it contained a rich reef. And so it did, being part of the greatest reef on earth.

The farm was later sold at a handsome price to Joseph Robinson, a brilliant but unlikable man, the second richest in Kimberley. But the richest, Cecil Rhodes, who knew little of gold mining, made the best deal. News was brought to him of events. Two Australian miners then demonstrated the quality of the finds to him the backyard of his cottage, and he persuaded the colleague who had brought him the news, Dr Hans Sauer, to return to the Witwatersrand and acquire interests for him, writing out a cheque for £200 as ready money. The doctor later said that the cheque was the most remunerative Rhodes ever signed 'for he amassed a far greater fortune out of the Witwatersrand than he ever did out of diamonds.'

Rhodes, who had stated back in 1877: 'I contend that we [the British] are the first race in the world and that the more of the world we inhabit the better it is for the human race', was now in a position to shape the economic development of South Africa, just as his great rival, Paul Kruger, President of the Transvaal since 1883, was to shape its political course. Kruger was to welcome *uitlanders* (foreigners, i.e. non-Boers) to develop the mines of the Witwatersrand, but was to deny them political rights and tax them heavily, two of the causes of the second Boer War.

Long before this, indeed soon after the original major finds of 1886, any semblance of true rush had vanished. Not that the capitalists of Kimberley who moved in so rapidly took up the whole of the great reef, for a few thousand prospectors tried their luck, breaking off pieces of exposed banket, pounding it up into mortar, then transferring the powder hopefully to their pans. If the results were good enough to stake a claim, the object of the exercise was to sell out as soon as possible, common sense perhaps, but hardly the stuff of adventure which was the trademark of diggers elsewhere. Ironically, the place where South African equivalents of the Argonauts flourished was on the floor of the Johannesburg Stock Exchange. Sometimes speculation was suspect, for we read how in May, 1887, with Johannesburg ready for a boom, 'Sensational crushings are being made from one rich narrow vein of banket for no other purpose than to off-load shares onto a still gullible public, and to refloat companies on the London market for amounts in no way warranted by the work done or the quantity of auriferous ground exposed'. So wrote E. P. Mathers in 1887 in his *The Goldfields Revisited*. But the liveliest account of the Stock Exchange was written by E. E. Kennedy in his *Waiting for the Boom*, describing the atmosphere in 1889. 'I.D.B.' is, of course, illicit diamond-buying, winked at by many ordinary citizens who felt that it at least gave the not-so-rich a chance of getting their own back at the capitalists who ruled the roost. This virtually marks our farewell to South Africa. The story of the Klondike which follows it is not so important a one in the history of gold production, but is far more significant in the human story of actual gold rushes, hence its greater length. But back to Kennedy:

> Among the members were men who had been storekeepers, canteen-keepers, lawyers, policemen, farmers, ostrich-feather dealers, clerks, bookmakers, one or two defaulting brokers from London, and there were some who were said to have been dealers in old clothes, and a good many of them looked as if it was their natural calling. There were men from Kimberley, too, some of whom were know to have taken the degree of I.D.B.

These members were dressed far more democratically than their opposite numbers in London:

> To a man fresh from the London Exchange, where an individual is chaffed for a whole day if he wears a loud necktie, a gaudy pair of trousers, or something special in waistcoats, where it would be simply seeking for destruction of the offensive article to walk in with any hat on your head but the time-honoured chimney-pot, the costumes of the Johannesburg Stock Exchange are a rude shock. These people wear every kind of headgear except the chimney-pot–helmets, deer stalkers, cricket caps, and even tam-o'shanters. The weather is cold in the early mornings, so there are many ulsters, some of remarkable design and colour; there are men in riding breeches and top boots, who carry a handsome crop and look as unlike stockbrokers as anything you could imagine.

The actual building which housed this marvellously clad array was soon too small to accommodate them all and a street market took the overflow, where hundreds of 'brokers, buyers, and sellers are always to be found ... the din of business bewildering to the passer-by'. Here also 'the larger part of the landed property for sale in the town is put up for auction'. Such was the report in a London paper of 1888, and it is worth noting that the membership of the stock exchange rapidly rose from 130 to just under 1,000. No wonder that the crowded building became the city's social centre, balls being held there, along with meetings of the Public Library and Sanitary committees!

A new stock exchange building was demanded and building got under way. Let it not be thought that trading was confined to office hours, for it went on in bars, hotels, indeed wherever anyone felt the urge to conduct business. Stock fever mounted as gold fever did in other areas, and soon Kimberley and many more cities and towns caught the shares mania. 'News will come down of some rich strike on the Rand,' related the *Diamond Fields Advertiser*, 'and at once the street will be crowded with brokers shouting out the prices, like bookies on the rails before the start of a big race.' Women were as much affected by the gambling mania as men.

The opening of the new stock exchange in 1888 was as spectacular an event as the coming of the railway in 1890, which by any standards was more important. Before trains reached Johannesburg, coaches served the public and there was great rivalry between the different firms. Certainly the first train in was greeted with rapture by Press and public alike, but the stock exchange, old or new, remained the heart of the city. Had not whole barrels of champagne and whisky been rolled in to it for parties to celebrate the first occasion that 10,000 ounces had been produced from the Witwatersrand in a month, February 1888? Every office block in the area was decorated with flags. And when the new building opened a cabinet minister came up from the Cape to perform the opening ceremony.

Yet perhaps it would be right to end this short account of the Golden City in its lusty heyday with a subject even more universally popular there than playing the market, the pubs (saloons), hundreds of them, with names which ranged from the Kentish Tavern to the Crystal Palace, the Welsh Harp to the Diggers' Arms. Many had back rooms where the proprietors made illegal fortunes selling drink to Blacks, but fortunes were to be made legally as well. Proof of the enormous popularity of the saloons was the Barmaids' Referendum of 1891, when 288 bars operated in a city of 30,000. A newspaper editor, Louis Goldsmid, decided to boost the circulation of his *Burlesque* by asking readers to nominate 'The most popular lady assistant in any bar, café or restaurant in Johannesburg' and 17,000 went to the polls. The contest went on for weeks until at last it was announced that Mrs Groth had come head of 20 contestants, with 1,956 votes, a personal triumph for herself and the Kimberley Bar. Second was Miss Birdie, also of the Kimberley Bar. What a party there must have been in the saloon that night.

An artist's impression of the road to the Transvaal gold-fields in 1887

The Klondike

At 3 o'clock this morning the steamer Portland from St Michael for Seattle, passed up the Sound with more than a ton of solid gold aboard. ...

Seattle *Post-Intelligencer*, July 17, 1897

It is the endless line of climbers on the four-mile trek to the summit of the Chilkoot Pass that has caught the imagination of successive generations, such is the power of photography. The famous pictures symbolise the unique quality of this last great gold rush to the Klondike, a remote, spectacularly rugged land where hot summers give way to cruelly low temperatures in winter, a land invaded by ill-prepared thousands within living memory.

Dominating this last frontier was the Yukon river, more than 2,000 miles long and rising a mere 15 miles from the Pacific, or, more exactly, the Gulf of Alaska to flow through Canadian, then American, territory to the Bering Strait. And though gold had earlier been found in Alaska, it was to be in Canada's Northwest near an obscure stream mispronounced 'Klondike' that the great strike occurred which sent men right across the world mad with a new disease, Klondicitis.

Alaska had been discovered by Russians in 1741 and was sold to the United States in 1867. Some gold had already been found, but the Russians were only interested in the fur trade, and Hudson's Bay traders, whose Fort Yukon was deep in American territory without their knowing it, were aware of the golden possibilities but were not greatly concerned.

Gradually, more and more Americans entered Alaska, some of them veterans of California and almost all of them experienced miners, not greenhorns, and from 1880 gold was found in modest quantities.

The miners, mostly American, also found some gold over the ill-defined Canadian border, and soon there were two 'towns' in the vast wilderness, Circle City in Alaska and Fortymile just across the Canadian border, though the latter was essentially an American outpost. The Canadian Government, anxious not to allow the area to become American simply because Americans were there, sent a 20-strong detachment of North-West Mounted Police under Inspector Constantine to Fortymile, the inspector acting as Agent as well.

This, as we shall see, was the beginning of another chapter in the matchless saga of the Mounties, who had been raised only 20 years earlier. 275 men had marched westwards to police what are now Saskatchewan and Alberta, drive out American whisky traders and protect the Indians, who trusted the men in red as much as they hated U.S. Army blue. When thousands of Sioux, among them Sitting Bull, fled across the border in the winter of 1876-77 after the killing of Custer, the legend of the Mounties really began, a legend based firmly on truth, for the story of how Major Walsh, alone, or with less than half a dozen men, went time and time again into Indian camps to confront warriors with the scalps of Custer's men joggling from their saddle horns is one of the great epics of the West. By the 1890s, the Mounties were world-famous, and the Klondike was to increase their fame and produce a veritable superman in Sam Steele, 'The Lion of the Yukon'.

But back in 1894, the region, with its few hundreds prospectors, was hardly even a frontier. There was not much crime even in the American Circle City, and at least one grizzled old miner was busily devouring Gibbon's *Decline and Fall of the Roman Empire*. 'Take 'em everywhere, and read 'em every night when I get time,' he told a visitor to his grubby shack. 'I'll bet I know more about Caesar, Hadrian, Attila, Belisarius and all the others than you do–or most anyone else.' And there had been one mining improvement which helped counteract the effects of the long winter. In the late 1880s, a miner had the idea of building a fire on the bottom of his frozen creek claim, and soon the practice of 'fire-setting' or 'drifting' caught on. This meant thawing ground every night and day until the bedrock had been reached, which might be 10-50 ft down in the sub-Arctic Yukon region. Then the miner could work along the 'pay streak' right through the winter in his burnt-out shaft. 'By morning,' noted W. B. Haskell in his *Two Years in the Klondike and Alaskan Goldfields*, 'if the amount of fuel has been properly gauged, nothing remains but the dying embers and the hot ashes; the smoke and gases have all escaped.' The pay dirt was piled in a mound known locally as a dump, and in the spring and summer the dumps were tested in the time-honoured ways, the pan, rocker and sluice box. At first the prospectors had optimistically hoped that the sun would thaw the earth, but wood fires were in general use before the historic day in 1896 when Bonanza Creek was found.

Only the remoteness of the Yukon can explain why it took so long for the Klondike Stampede to occur. As early as 1804, one Baranov, the old Russian governor at Sitka, saw a hunter draw a handful of golden nuggets from his pocket and, according to Jack London, writing in the *Atlantic Monthly Magazine* in July, 1903, uttered these striking words:

> 'Ivan, I forbid you to go farther in this undertaking. Not a word about this, or we are all undone. Let the Americans and the English know that we have gold in these mountains, then we are all ruined. They will rush in on us in their thousands, and crowd us to the wall–to the death.'

In the event, the Russians had departed long before history was made by two Indians and a white man, as a result of which 'sturdy, indomitable gold-hunters of Anglo-Saxon stock', as Jack London called them, headed for a place about which few of them knew anything.

Some, including the unfortunate miner himself, claimed that history was made by a dour, lanky Nova Scotian of Scottish descent named Robert Henderson. Grubstaked by a shopkeeper of French Huguenot descent, Joseph Ladue, he set out on a journey dogged by ill-fortune and disasters until he reached a mountain which he climbed to gaze in awe at the majestic peaks all around him.

Below him were a number of creeks, some of the richest gold-bearing creeks on earth. He panned in one of them, found eight cents worth, and hopefully christened the creek Gold Bottom, then went to a spot where some 20 miners were working on sandbars and persuaded three of them to come back with him. By the end of July they had taken out some $750 worth, but were almost out of food, so Henderson returned to Ladue's trading post, which was located at Sixtymile on the Yukon, telling everyone he met of the V-shaped valley in the hills where there was gold. Soon everyone but Ladue was heading for the spot.

Replenished with supplies, and with the Indian River down which he had come now shallow in high summer, he set off down the Yukon, sure that his creek was a tributary of the Thron-diuck, which itself was a tributary of the Yukon. Miners were already mispronouncing Thron-diuck 'Klondike'. The Indian word meant 'Hammer-Water' because the Indians hammered stakes across its mouth over which they spread their nets. The catch was salmon, which teemed in the river. The northern bank was swampy and covered with scrub timber, though hills rose to the north. It was an unlikely spot to house a world famous city, but here Dawson was to arise from nothing.

But all Henderson could see that day was a white man, and he pitied him for fishing when he might be striking it rich, and decided to tell him how to make his fortune.

The white man was George Washington Carmack, born in California, now married to a Siwash Indian girl called Kate, and with precious little ambition, having wed a chief's daughter, except to be a chief himself. By any standards he was an oddity amongst whites who looked down on the Indians, and he even liked being called Siwash

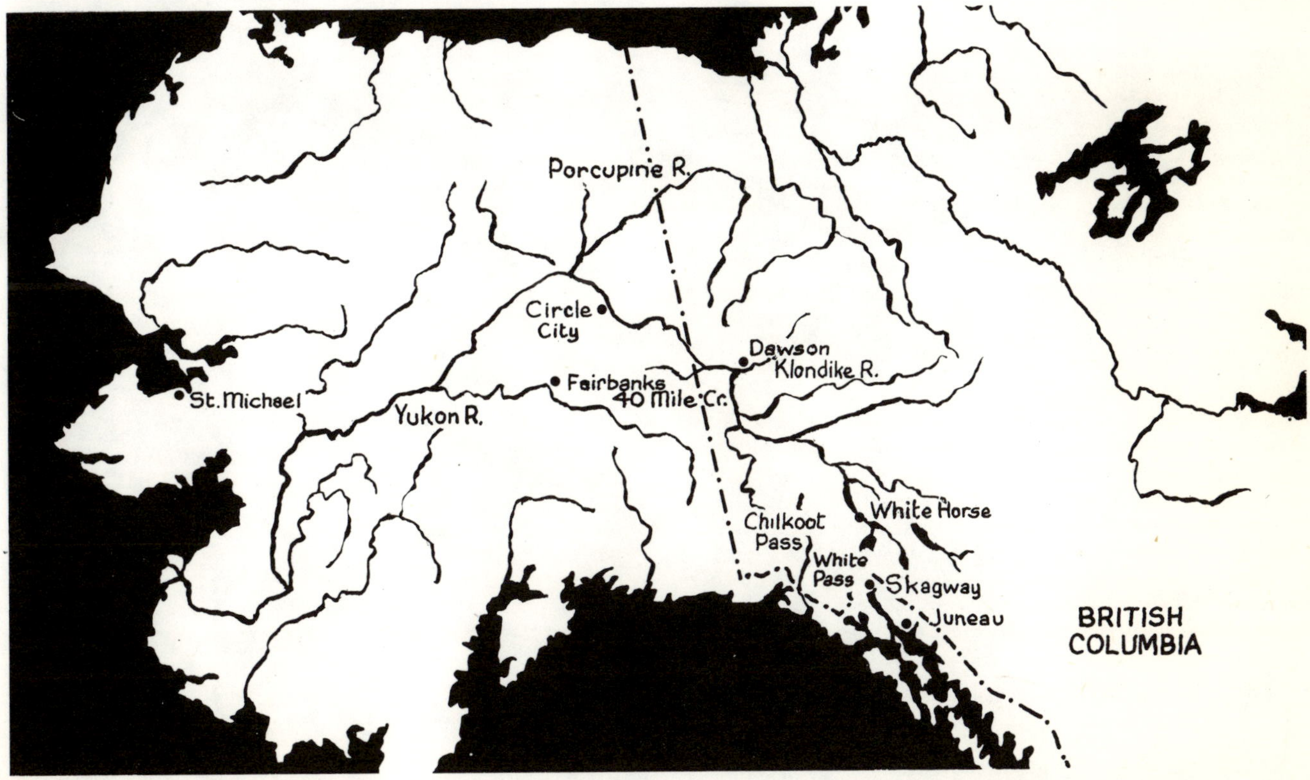

The Klondike, the Yukon and Alaska

The San Francisco Call
PRICE FIVE CENTS.
SAN FRANCISCO, TUESDAY MORNING, JULY 20, 1897.
VOLUME LXXXII.—NO. 50.
INEXHAUSTIBLE RICHES OF THE NORTHERN EL DORADO.
Official Verification of the Wonderful Gold Discoveries in the Klondyke District—Ten Million Dollars a Conservative Estimate of What Will Be Taken Out This Year.
PORTLAND'S OPPORTUNITY.
Oregon's Metropolis May Supply the 50,000 Who Will Rush to the North.
Going From New York.
AN "OFFICIAL" REPORT.
Dominion Surveyor Ogilvie Has Sent an Account of the Discoveries to Ottawa.
CHEST IN THE WORLD.
GOING FROM BERKELEY.
Party of Ten Men Being Organized at the College Town for Klondyke.

George. His tribe, unlike some in the far north, were easy-going, and so was Carmack.

Not that Carmack had cut himself off from white culture. He had an organ in his cabin as well as scientific journals and wrote poetry on occasion. As he stood waiting to speak to Henderson, Skookum Jim, tall, handsome and a famous local trapper and hunter, joined him, also Tagish Charley, 'as alert as a weasel', according to Carmack, chubby Kate, and their daughter.

Henderson told Carmack the good news, then proceeded to lose himself a fortune. Though his exact words are disputed, he probably said: 'There's a chance for you George, but I don't want any damn Siwashes staking on that creek.'

He pushed out his boat and left the group, returning to Gold Creek Bottom. Carmack had not taken his final remark well, and Skookum Jim–his first name meant 'strong'–was not amused either. But Carmack told him not to worry: they would go and find themselves their own creek. However, the trio, in no particular hurry, visited Henderson at his creek, but were not over-impressed and, after following it up into the hills, crossed down to Rabbit Creek, soon to be re-christened Bonanza Creek.

Yet it was not as simple as that, for something had happened between Henderson and Carmack when they remet at Gold Bottom. On the way there, the three had panned a little in Rabbit Creek, the tributary of the Klondike up which they had come and which emptied out opposite where Dawson was to stand. This was to be the fabulous Bonanza Creek where they would make their great discovery. This time they struck a little colour, but not enough to halt them.

According to Carmack's story, when he met Henderson at Gold Bottom, he told him to come over to Rabbit Creek and stake a claim, but Henderson was to swear that it was he who urged Carmack to prospect Rabbit Creek.

Whatever the facts, the anti-Indian Henderson ruined his chances once more, for though Carmack seems to have promised that he would let him know if anything worthwhile occurred at Rabbit Creek, when the Indians tried to buy tobacco from the gloomy Nova Scotian, he refused them.

Up Gold Bottom Creek, over the hills and down to Rabbit Creek went the trio, and on August 16, some half a mile from its junction with Eldorado Creek, they made history. Carmack always claimed *he* made it, happening upon a protruding rim of bedrock and extracting from it a chunk of gold the size of a thumb. Skookum Jim and Tagish Charley told a different, probably correct, story, that Carmack was sleeping under a birch at the time and Jim, after shooting a moose, was cleaning his pan in the creek and found the gold.

The birth of Klondicitis, a disease that was to spread round the world

The result was the same, for there undeniably was the gold, sandwiched between slabs of rock, and a single pan producing a quarter of an ounce, worth some $4, a major find in the Yukon, where 10 cents worth was a cause for celebration.

A war dance ensued, later described by Carmack as a combination of Scottish hornpipe, Indian fox trot, syncopated Irish jig, and Siwash hula, then they rested, smoked and set to work again. And as that marvellous day ended, so did Carmack's time as an Indian. Siwash George was transformed into George Carmack, would-be gentleman of means.

The next day four claims were staked, two for Carmack as the discoverer, as allowed under Canadian law. Skookum Jim later said that this was agreed because, as an Indian, his claim to be the discoverer would not have been recognised. Carmack cut away enough bark from a spruce to be able to write in pencil:

TO WHOM IT MAY CONCERN

I do, this day, locate and claim, by right of discovery, five hundred feet, running up stream from this notice. Located this 17th day of August, 1896.

G. W. Carmack

They recorded their find at Fort Cudahy on the Yukon, then went to nearby Fortymile. On the way from their claim they had told everyone they met, and doubters were quickly transformed by a sight of the gold. It was to be the same at Fortymile, but they sent no word to Henderson. In the main saloon at Fortymile, the squawman's story was not believed at first. Some knew him as 'Lying' George because he had once (correctly) claimed to have found coal and had always tried to suggest that things were going better for him than they were. But now? That night men started to slip away to Rabbit Creek. After all, they had seen George's gold, and perhaps he really had found it. A fortnight later Bonanza Creek was staked out.

Despite the excitement, many veterans of other rushes and disappointments were at first suspicious of this new wonder strike, especially at Circle City. It needed reports of a $65 pan, then of one worth $212, to cause a stampede, which, as it happened in midwinter, was rugged in the extreme.

One man who had no doubts was Henderson's grubstaker, Joseph Ladue, who laid out the townsite of Dawson in September, naming it after a Government geologist. Meanwhile, the unfortunate Henderson heard nothing of the strike until early September, when a group of miners told him of gold at Bonanza Creek, a name new to him. When they pointed out where it was, his heart must have

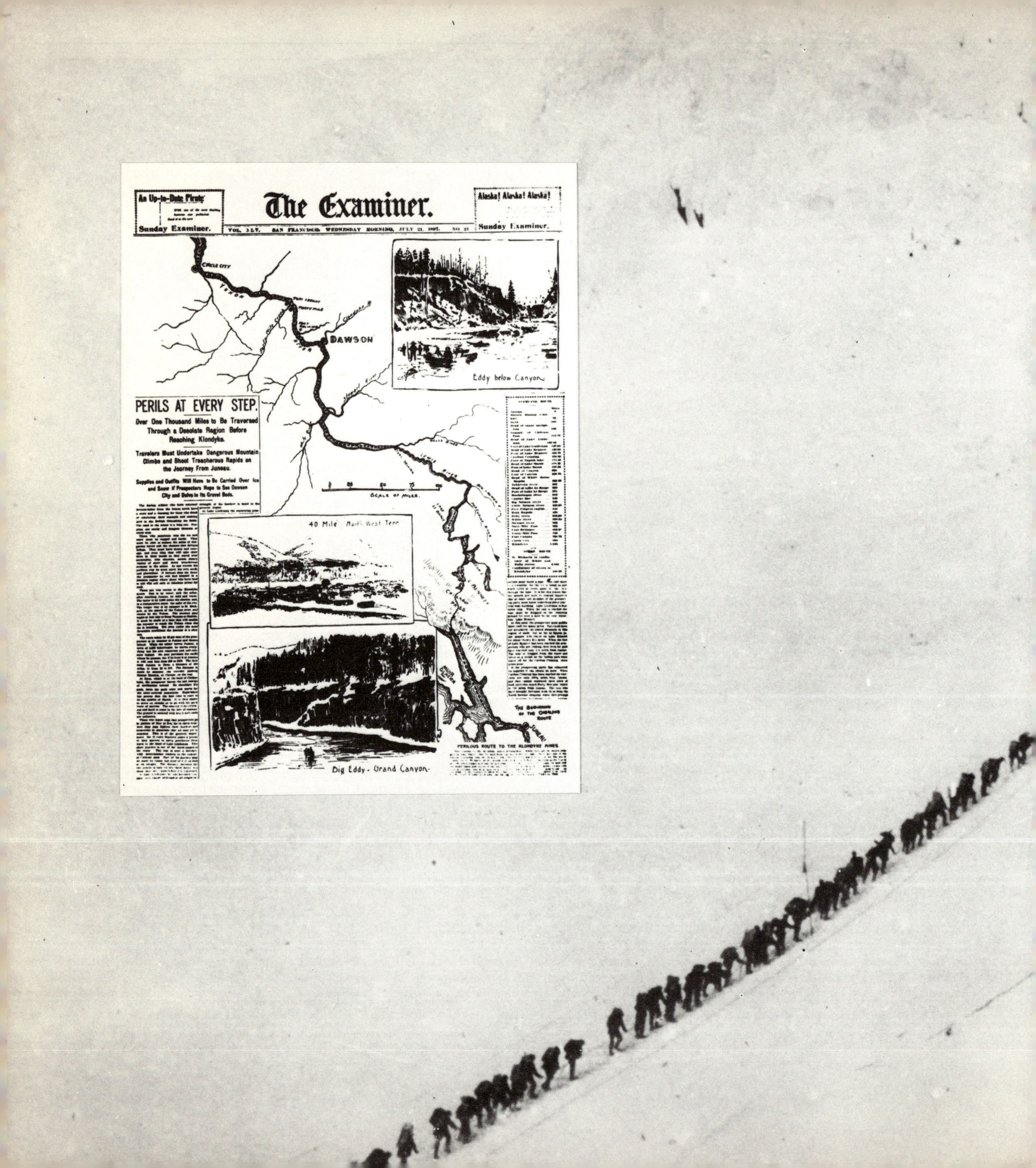

An Up-to-Date Pirate

Sunday Examiner.

The Examiner.

VOL. XLV. SAN FRANCISCO, WEDNESDAY MORNING, JULY 21, 1897. NO. 21

Alaska! Alaska! Alaska!

Sunday Examiner.

PERILS AT EVERY STEP.

Over One Thousand Miles to Be Traversed Through a Desolate Region Before Reaching Klondyke.

Travelers Must Undertake Dangerous Mountain Climbs and Shoot Treacherous Rapids on the Journey From Juneau.

Supplies and Outfits Will Have to Be Carried Over Ice and Snow If Prospectors Hope to See Dawson City and Delve in Its Gravel Beds.

Eddy below Canyon

40 Mile. North West Terr.

Big Eddy. Grand Canyon.

PERILOUS ROUTE TO THE KLONDYKE MINES.

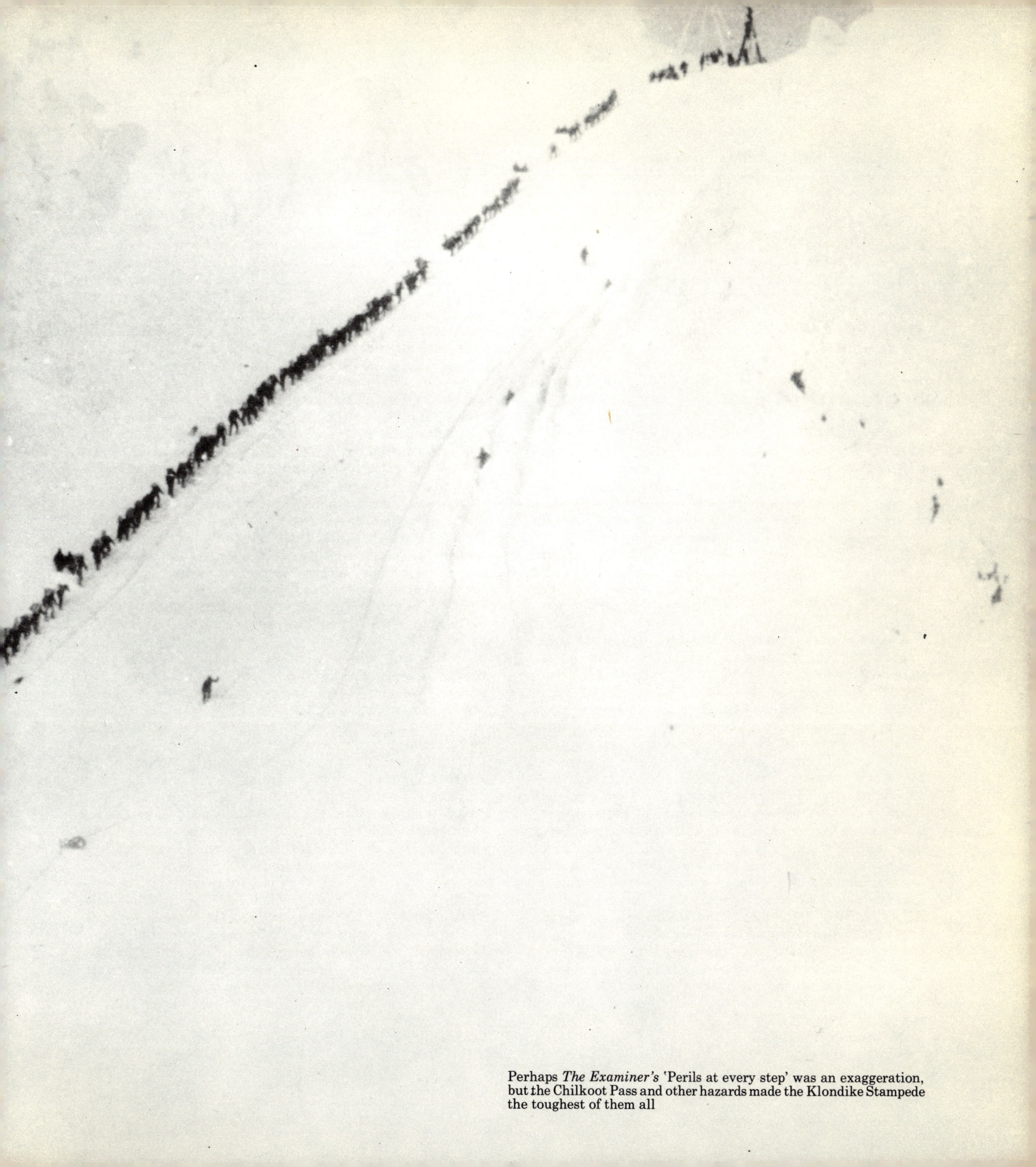

Perhaps *The Examiner's* 'Perils at every step' was an exaggeration, but the Chilkoot Pass and other hazards made the Klondike Stampede the toughest of them all sunk, then he heard who had found the gold, flung down his shovel, walked to the bank of his Gold Bottom Creek and sat down, his life blighted. When he reached Bonanza Creek he could only obtain claims of slight value, or so they seemed, yet one of them, after he had sold it to pay medical bills, proved enormously rich. On the boat back to Seattle in 1897, all he had made was stolen. Though Canadians and British regard him as the truest finder of Klondike gold–the Americans support Carmack–it was not for many years that he was given a reward, a pension of $200 a month. He went on looking for his bonanza all his life, dying in 1933, but his son continued the quest, hoping to find the mother lode of the Klondike, which probably vanished long ago, destroyed by glaciation and erosion. He, too, died a frustrated man.

The famous three of Bonanza Creek were more fortunate, though Tagish Charley's capacity for drink was the ruin of him. Made a Canadian citizen, like Skookum Jim, he sold out in 1901 and opened a hotel at Carcross, where he became a splendid host and an honorary white man, and therefore was allowed to drink too much. One day, when drunk, he fell off a bridge and drowned.

Skookum Jim was made of sterner stuff. Despite getting $90,000 a year from his mine, he went on prospecting, endlessly searching for a quartz lode that he never found, and so taxed his splendid physique that he died from sheer overwork in 1916.

As for Carmack, rich beyond his dreams, he and his Kate visited the Pacific Coast of the U.S.A., their hobbies including tossing money out of their hotel windows in Seattle and San Francisco and enjoying the excitement below. To make sure she got back to their room safely, she used to blaze a trail with her fish-skinning knife, nicking the wood in whichever hotel they were staying. But in 1900, Carmack sent her back north, for she was never really happy, landing in jail after drunken sprees and fights. She returned to her old home with her half-blood children, lived on her government pension, never without her necklace of nuggets from Bonanza Creek. She died in 1917.

Her George married again soon after Kate had gone, transferring his affections to a lady named Marguerite Laimee, who had served the needs of miners in South Africa, Australia and finally the Yukon, where she ran a 'cigar store' in Dawson. This did not automatically mean bawdy house, but she was known to clear some $30 each morning panning the sawdust of her floor for gold dust. The happy pair settled down, and Carmack, when he died in 1922 in Vancouver, also had a Californian mine and ran an apartment house and a hotel. His wife, inheriting his fortune, died in 1949.

The world knew nothing of the Klondike that first winter of 1896-97. Already the pattern of law and order–which did not mean that high jinks were banished–had been set. Because the original staking on Bonanza Creek had been inefficient, disputes arose and sometimes claimants slugged it out, but in January, William Ogilvie, a government surveyor, reached the Klondike and untangled the confusion, first insisting that his decisions must be considered final. It speaks volumes for him that they were, considering that a man stood to gain or lose a fortune over a matter of a few feet. The presence of Inspector Constantine and his Mounties almost from the start, ensured the rule of law. Neither he nor Ogilvie, nor the latter's son Morley, made a cent from the strike, though Constantine allowed his men to stake claims, and some struck it rich. So the richest spot in the world that winter, though the world at large knew nothing of it, was ruled by two men.

If anything, Ogilvie improved on his own high standards. He had to, for every inch became important. Incredible riches were found where Bonanza and Eldorado creeks met. $1,000 came from one pan there, and on the next claim, owned by an ex-Black Hills miner named Dick Lowe, nuggets were apparently visible at 20 feet. Inevitably, he lost some gold to thieves, for it would have needed a constant armed guard to protect his property, but his own riches were enough for any man, even though he was one of those whose fortune turned him into a permanent drunk.

So the long winter went on, and the Klondike even produced a king, one Big Alex McDonald, who had mined silver in Colorado and now decided it was easier to buy up property than dig for gold himself. Towards the end of 1897, the King of the Klondike held interests in 28 claims and was thought to be worth millions.

Meanwhile, Dawson was growing slowly, as news of the strike brought miners to it from all over Alaska. The first of many thousands climbed the Chilkoot Pass and built boats along the shores of Lake Bennett waiting for the ice to crack to sail down the Yukon, which rose in the lake, to the gold-fields. But many of these newcomers were heading for Circle City and knew nothing of the Klondike, and when the ice broke in May and they sailed down river, they were startled to see two towns, Klondike City on one bank, usually known as Lousetown because the Indians had once had a salmon camp there, and Dawson on the other. Lousetown could have become the main centre for the Klondike if its two saloonkeepers had had their way, but they were foiled by Ogilvie, who had a nice sense of humour. He tied down the steam whistle on Joseph Ladue's sawmill, which acted as a piercing sirens' song to the little boats coming down the river, and most of them sailed by

Klondike Stampeders waiting for the ice to break at Lake Bennett in 1897–8 before sailing down to Dawson the irate citizens of Lousetown and berthed at Dawson, where by the end of the summer ten saloons were doing roaring business, transactions in them, and, indeed, everywhere else in town, being mainly in gold.

Even the floors were panned for gold. In Big Harry Ash's Northern Saloon, the proprietor offered a boy $25 for all he could extract from the sawdust in front of the bar and the result was $278 worth. It came in fine dust that sifted through the pokes–the little bags–in which miners carried their gold and which they slapped down on the bar.

The first steamboat, the small sternwheeler *Alice*, docked at Dawson after a 1,700-mile journey up the Yukon, in June, bringing food and drink in equal proportions, and two days later, another vessel, the *Portus B. Weare*, arrived, giving the town another excuse for a spree only hours after it had sobered up from the first one in which all the saloons dispensed free booze.

Those who had already struck it rich had been amongst the most eager for a sight of the *Alice*, for they wanted to get to the outside world as quickly as possible, some to enjoy themselves as never before, then return for more treasure-hunting, others, the sensible ones, to live securely off what they had already made. There were over 80 of these Bonanza and Eldorado kings, whose return, like so many Drakes aboard *Golden Hinds*, was to set off the last great gold rush, and their fortunes ranged from $25,000 to $500,000. Their occupations in earlier existences varied from blacksmith to bookseller, from lecturer to loafer, and now their only problem was how to stow all their gold aboard the two small vessels which would transport them to St Michael on the coast where both gold and owners would transfer to ocean-going craft.

The gold travelled in boxes, cases, jam jars, strapped up blankets, any available bottle, packing cases and in pockets. The *Alice's* decks groaned under the weight of the dust and nuggets. Cabins, already on the small side, were smaller still when the gold was shipped aboard, while the three-storied *Weare* had to have its decks shored up with wooden props to prevent collapses. The victors bearing their spoils down river could not live it up aboard their treasure ships, for food was in short supply. The first feast came at St Michael, where at last they could enjoy the fresh fruit and vegetables they so craved. Then it was time to go aboard the *Portland*, Seattle-bound, or the *Excelsior*, whose destination was the city that had seen it all before, San Francisco. But it had never seen a whole boatload of Bonanza and Eldorado kings.

Everything conspired to make the arrival of the two ships, bearing their thrilling news, a world-wide sensation. America, with its population explosion, was running short of gold that summer of '97, just under a half century since James Marshall's find, while the Pacific Northwest was actually experiencing a slump. The Nineties were reaching their frenzied climax of feverish revelry and considerable misery in an age when millionaire often meant robber baron. The four 'grocers' of the Californian gold rush had turned to railroads and become richer than ever, the most famous of them, Leland Stanford, dying in 1893. Gold was as emotive a word as ever, and now communications were infinitely more rapid, so that Britain in the high noon of Imperialism would hear the news of the Klondike when America did and young men would hear the call of the wild and exult that it had happened in part of the Empire. In fact, the Klondike was in terms of gold extracted, to prove one of the smaller rushes, but it was, and it remains, one of the most famous of them all. The world was ready for a sensation–the Spanish-American War and the Boer War had not yet occurred–and there was even a sensational press ready to increase the fever. In terms of impact, the Klondike was the most internationally stirring of the gold rushes and Klondicitis was a very real disease.

The *Excelsior* arrived first, docking at San Francisco on July 15, her decks filled with mud-stained working men who happened to have made fortunes. Many drove straight to the Selby Smelting Works, the official mint being closed because of political changes, and stories of how the gold was scooped into the melting pot were soon on every lip. True, there had been rumours of a big strike before this day of days, and Ogilvie's report had already been printed, but the physical fact of the ship and its human and mineral cargo caused the explosion of interest, increased when the *Portland* docked two days later in Seattle. 'Show us the gold,' roared thousands on the waterfront, and the Bonanza and Eldorado kings lifted their sacks. When they landed, they were welcomed, like their San Francisco counterparts, more frenziedly than modern pop stars, and some became almost crazed by it all, others prostrated with exhaustion, and others a combination of both.

Rumour soon multiplied the finds and the Seattle Chamber of Commerce rose to the occasion, the city, according to a New York reporter, having 'gone stark, staring mad on gold.' The city fathers retained their sanity enough to ensure that their charge became the centre of the gold rush trade–excepting those who headed for the Klondike themselves! Seattle above all was to be denuded in the way San Francisco had been so many years before.

Only a small percentage of those who joined the great stampede northwards were miners, and the numbers of greenhorns reached record proportions. Many were totally unfitted for the rigours of the Klondike of which they knew so little when they set out. Shopkeepers did their often dishonest best to supply their wants, adding the magic word 'Klondike' to innumerable goods on display. The majority of the stampeders were American and Canadian, but every part of the British Empire where gold had been found sent its quota, as did the Mother Country. As usual, there were South Americans eager to head for the gold-fields, and the old Californian rush designation 'Argonauts' was heard once again.

Sophisticated travellers bought up evaporated food–eggs, soup etc.–which was all the rage, along with milk tablets, saccharine and other items unknown to earlier adventurers, while Tappan Adney, representing *Harper's Illustrated*, met a man in Victoria who was travelling with 32 pairs of moccasins, plenty of ordinary shoes, also pipes, two Irish setters, a bull-pup and a tennis set. He, apparently, was merely going for a good time, not to find gold.

Gradually, the world discovered roughly where the Klondike was, though how to get there remained a problem. A Londoner asked an old Alaska hand if one could bicycle over the Chilkoot, whilst at least one would-be prospector asked the Seattle station master which train he should catch. Favourite slogans of the day were 'Ho! For the Klondike' and 'Klondike or bust!' and publishers were soon rushing books, pamphlets and every other sort of literature into print about the longed-for spot, much of it as ill-informed, though not so deliberately, as the rubbish which afflicted the Pike's Peak rush.

Amid all the excitement, a few dismal voices were raised advising caution. But who wanted to listen to the sensible statements of Louis Sloss of the Alaska Commercial Company:

> I regard it as a crime for any transportation company to encourage men to go to the Yukon this fall.... The Seattle people who are booming the steamship lines may be sincere, but a heavy responsibility will rest on their shoulders should starvation and crime prevail in Dawson next winter.... It is a crime to encourage this rush, which can only lead to disaster for three quarters of the new arrivals.

Thanks to the Mounties there was to be neither starvation nor crime in Dawson, but otherwise Mr Sloss's remarks were accurate to a degree. Nobody listened. And neither did they listen to the stern warnings that after September there was no chance of reaching the Klondike until the following summer. Already, despite official pleas, some 9,000 were on their way and, with another boat-load of gold in San Francisco, the thunderous urgings of both newspapers and vested interests, there was no hope of holding back the flood until the spring of '98. From Victoria southwards every city's merchants were screaming the advantages of their wares, ginger being

Snake-hips Lulu, a popular attraction in gold rush Dawson

added to the argument by trade rivalry between Canada and America.

There were four main routes to the Klondike. Two of them, the long journey up the Yukon and the shorter but terrible trek over the adjacent Chilkoot and White passes, were the more used. The other two were the Ashcroft Trail north through British Columbia, a nightmare trip, and the Edmonton Trail which was a killer.

The Ashcroft began at the town that gave it its name, 125 miles northeast of Vancouver, then headed through country the British Columbian miners had known and on into a wilderness of bogs, forests, rivers and endless rain, horses and men being continually under attack from flies and mosquitoes. By the summer of '98, the trail had been denuded of fodder and horses died by the score. The forests became steadily thicker, and latecomers came across spots with names like Starvation Camp and Poison Mountain. Many gave up, others continued their hellish 1,000-mile trek, linking up with still more unfortunates at Glenora, who had fought their way up the slush and ice of the Stikine River, a route publicised by the British Columbia Board of Trade as a way of avoiding the dangers and hardships of the passes etc.

Early hydraulic mining in California

Many of them had suffered at the hands of Soapy Smith's gang, who had already spread out from their headquarters at Skagway, of which more anon, to the town of Wrangell at the mouth of the river. Some 1,500 stampeders tried these two routes in '97 and '98, but only a handful ever reached Dawson.

Far worse, however, was the Edmonton 'route'. This was promoted by the city's Board of Trade as the perfect route, one which would take a man only 90 days with the aid of horses, Indian canoes and Hudson's Bay Company supplies *en route*. In fact, this 2,500-mile trip along the Mackenzie River almost to the Arctic Circle, or the 'short' 1,700-mile trip via the Peace River were appalling experiences for frontiersmen, and for the majority who used them, rank amateurs, akin to being sentenced to death. The statistics prove this is no exaggeration, for of the 2,000 or so who set out from Edmonton, at least 500 died. Superintendent Sam Steele of the Mounties declared that it was 'incomprehensible that sane men' should risk these routes, but, sadly, gold fever has always caused rosy claims and advertisements to be swallowed whole. By late 1899, perhaps 100 had reached Dawson, by which time the gold rush was over. Some had been travelling for two years. Behind them was the smell of dead horses, a stench which permeated all the routes to the gold-fields, even the possible ones.

Meanwhile, the vast bulk of the stampeders had taken sane, if hardly comfortable, routes. There was a serious shortage of ships, with vessels called into service which were just able to float, but grossly ill-suited to the conditions they would meet. Many captains had never sailed on an ocean, being old rivermen, and their crews were often no better.

Overcrowding was inevitable. One of the 20 ships that sailed for Alaska in the first five weeks of the stampede was the *Al-ki*, with not an inch to spare. A thousand had hoped to sail in her and had besieged the dock, not daring to leave for a meal, but finally the tiny ship set out with 110 humans, 900 sheep, 65 cattle, 30 horses and 350 tons of supplies. The *Amur* was 'a floating bedlam, pandemonium let loose, the Black Hole of Calcutta in an Arctic setting', with 500 instead of 100 passengers aboard and an army of big dogs, all of them to prove useless in the Klondike. Passengers, who included 50 prostitutes, were packed 10 to a cabin, though tickets had promised them a cabin each and meals took a full seven hours to serve in the tiny dining cabin. Stewards rarely got from galley to table unscathed, for frantic travellers seized what they could *en route*.

It was 1,700 miles upstream to the gold-fields, but, more significantly, the great river was frozen solid by late September. Many spent the winter at Fort Yukon or Circle City, joined by others who had fled from Dawson, men who had managed to reach the Klondike in that autumn of 1897 only to find that food supplies were dangerously low. That they left was a stroke of fortune for those newcomers and oldtimers that remained, for though starvation threatened Dawson that winter, the worst never happened.

But along the Yukon in Alaska despair often turned to rage, as hungry, bitter men, denied access to their goal, or driven from it in alarm, sometimes turned to violence in remote places where law and order were virtually non-existent. One unfortunate government employee, Captain Ray, an army officer sent to examine ways of helping destitute miners, wrote: 'The tide of lawlessness is rising rapidly along the river.' And through no fault of his or of his only subordinate, it continued.

This was lawlessness essentially unorganised, and almost inevitable. In Skagway, the gateway to the Chilkoot and White passes, things were very organised indeed. But the passes were in use before organised crime on a monumental scale gripped Skagway, so they must be described first.

From Skagway, and Dyea three miles beyond it, to Dawson was a journey of 600 miles. The White Pass, 600 feet lower than the Chilkoot, was a more roundabout route to Lake Bennett and the headwaters of the Yukon, and few of the 5,000 who used it in 1897 reached the river before it froze for the winter. Every sort of physical difficulty was encountered from swamps to a path on Devil's Hill only two feet wide with a drop of 500 feet below it, and Summit Hill down which poured rivers of mud.

But the White is remembered mainly for the terrible fate of the horses. Many had been bought for high prices in Seattle and Victoria, though only fit for the glue factory; some were unbroken; most of them grossly overloaded by men who knew nothing of horses. 'The horses died like mosquitoes in the first frost and from Skagway to Bennett they rotted in heaps,' wrote Jack London, and Major Walsh, now retired and on his way to the Klondike as Commissioner of the Yukon, wrote to his superior in Ottawa that 'such a scene of havoc and destruction' could scarcely be imagined.

The story of the White Pass shamed the Klondike Stampede, but not enough to tarnish the inspiring courage and determination of those who went over the Chilkoot, incredibly some 22,000 that winter when the White Pass ceased to function.

The Chilkoot, for all its four-mile climb to the summit, was a short cut to Lake Bennett. It was too steep for horses, so the endless chain of humans bent nearly double by their loads, wound their way up the pass towards the summit. Because of the threat of famine in the Klondike, the Mounties, posted at the head of the pass, demanded that each stampeder brought in a year's supplies–food, and essentials like a tent and tools–which meant more than one journey up the Chilkoot. 'Helpers' in the shape of Soapy Smith's gang were there, though they never got past the Mounties and into Dawson. Which is a good moment to meet Jefferson Randolph 'Soapy' Smith, the uncrowned king of Skagway, Dyea and other Alaskan towns in the approaches to the Klondike, but especially of Skagway.

Smith reached Skagway in the autumn of '97, having earlier announced in Seattle that he was going north to become its boss. It was already a notably wicked spot, 'the roughest place in the world', according to a later visitor, Superintendent Sam Steele, who would be taking over from the admirable Inspector Constantine in Dawson in 1898 to become the Hero of the Klondike as much as Soapy Smith, denied access to it, was its Villain. In his *In Search of Eldorado*, Alexander Macdonald gave a Briton's view of the town as it was in late 1897:

> I have stumbled upon a few tough corners of the globe during my wanderings beyond the outposts of civilisation, but I think the most outrageously lawless quarter I ever struck was Skagway ... It seemed as if the scum of the earth had hastened here to fleece and rob, or ... to murder ... There was no law whatsoever; might was right, the dead shot only was immune to danger.

Smith did not appear to be scum, for he looked almost as respectable as one of his leading henchmen, 'Reverend' Bowers, of whom it was said that he exuded piety from every pore. Soapy's extraordinary nickname was not bestowed at birth–In Georgia in 1860–nor when he was a young cowhand and decided that a con man's life might suit him more. It was bestowed on him by a policeman who booked him in Denver.

He had been practising his most famous con trick, learnt from an old man in Leadville. The trick was to convince miners that the $5 cakes of soap he was selling on the street might have dollars under their blue wrappers. The first one bought had a $100 bill wrapped around it–the crowd had watched the soap being wrapped and Soapy made sure that the bill was seen–but nearly all the rest were simply–soap.

The act was helped by Soapy's style. He would begin by insulting the miners for their undeniable dirtiness, but this could be cured, as he told them in his admirable voice, which oozed sincerity, by 'the finest soap in the world, perfected and manufactured in my own factory.' It was a marvellous act.

The Law had better things to do in the West than worry about con men, who were often in league with the police. Usually, though, the fraternity were moved on by an officer, but Jeff Smith used to stand firm. He wanted to be known, and, instead of standing at a street corner like lesser artists, worked from a wagon drawn by a fine team of horses. One day he was picked up and booked, but when the lawman came to write the minor case up, he could not remember Smith's first name, so wrote 'Soapy' in brackets. It was as simple as that.

An ambitious man, Smith rapidly became king of Denver's underworld with an army of con men under his command, all devoted to extracting money from visitors–locals were usually and tactfully left alone–and seeing that a sucker never got an even break. 'Reverend' Bowers was one of his finest assistants, having a genius for warning a newcomer of the temptations of the wicked city, then leading him, if he really was determined to gamble, to men he could really trust. That was usually the end of his bankroll.

Naturally, Soapy's gang began to use violence to help oil their operations, but Soapy was rarely tarnished, protecting his image carefully and giving to charity as often as possible. He was his country's first great modern racketeer, and some at least of his good works appear to have been done spontaneously. Suckers, however, were up against the golden rule: do him good–and plenty.

If anything, his own soap act got better, with one of his gang making sure he got the first prize, then letting out yells of glee and rushing off with his $100 or $50 bill, thus causing a rush towards Soapy in case the prizes ran out. In private life Soapy lived like the perfect husband and father, another anticipation of 20th century underworld kings. Reforming politicians tried to oust him, but he was the ultimate born survivor, and when Denver finally proved to hot for him, he ruled in Creede. That he was a crack shot with a six-shooter in emergencies was an extra bonus, but finally Colorado had seen enough of him and 'Colonel' Smith–a self-appointed rank–decided to head for the last frontier.

Skagway was Soapy's masterpiece. His gang was initially seven, but rapidly increased to a hundred and it became virtually impossible for a stampeder to get through the town without falling into its clutches. The gang had been short of money when finally they were forced out of Colorado, but they were not short of money now. Never mock at a Hollywood scriptwriter who shows a town totally in one man's pocket. In a matter of weeks Soapy owned Skagway.

It helped that so many of his boys looked not only innocent but eminently respectable from their bowler hats downwards. Amongst the cream of con men were three as useful as the Reverend. On hand when the ships arrived was a newspaperman, Billie Saportas, who had arrived legitimate, but was given a job on *The Alaskan*, whose editor was on Soapy's payroll. Bill interviewed newcomers and found out how rich they were. 'Slim Jim' Foster was at the docks also, carrying bags uptown for the new arrivals and checking up still further on them, while particularly impressive was 'Old Man Tripp', complete with long white hair and an air of integrity. His act was that of returning stampeder and he was to be found on the trails, dispensing false information as he checked on who had any money left after escaping from Skagway.

In town Soapy had a number of sham enterprises to which suckers could be usefully steered; a Telegraph Office, a Cut Rate Ticket Office, a Reliable Packers, an Information Bureau and a Merchants' Exchange. Slim Jim's natural rendezvous was the Reliable Packers. What more could a stranger want than an 'honest outfit' to carry gear over the pass without overcharging?

At all these establishments it was traditional to demand a small deposit to ensure that business would not be taken elsewhere, or so the sucker was told, though actually it was to get a glimpse of his wallet. If it was suitably loaded a notable display of histrionics occurred, various members of the gang staging a convincing scene, starting with a low-looking type seizing the wallet, another man, a picture of honesty, rising in righteous wrath at the outrage, and others rushing hither and thither. In the confusion, the victim might or might not be floored, but one thing was certain: his wallet was lost for ever. One of the few men to get his money back from a Skagway robbery was Bishop Rowe, the pioneer bishop of Alaska, who lost a pouch of gold to one of the gang. But when the robber found out who his victim was he returned the gold.

'Why do you give this back?' asked Rowe.

'Hell, Bishop,' was the answer, 'I'm a member of your congregation.'

As he had in Colorado, Soapy kept on the right side of ordinary citizens in Skagway, making sure that they were never robbed, and subscribing to local charities, but in the end, like Henry Plummer, his sins caught up with him. Though we shall see his men at work on the Chilkoot later in this chapter, it is high time to go forward to July, '98 and remove the arch criminal from the limelight he so enjoyed.

By now many citizens, even those who could not help liking Soapy, despite his patent villainy, were becoming worried by Skagway's reputation. The town was the obvious shipping point from the gold-fields, but with bandits in control how long would people tolerate it? Skagway might become a ghost town. And there were citizens of the best type in Skagway, who had been working towards a situation when public opinion would turn against the gang. Above all there was Frank Reid, a surveyor from Dyea, who had been mainly responsible for organising a vigilante committee which as yet had no teeth. Its first proclamation sounded fine, but was no threat to Soapy's iron control:

WARNING

A word to the wise should be sufficient. All confidence sharks, bunco men, sure-fire men, and all other objectionable characters are notified to leave Skagway and the White Pass. Failure to comply with this warning will be followed by prompt action!

Signed–COMMITTEE OF ONE HUNDRED AND ONE

Soapy promptly started a Committee of 303.

But Frank Reid worried Soapy, one of the only two men in Alaska and the Yukon that he feared. The other was Sam Steele, of whom Soapy said: 'It's men like him that make it a bum place to live.'

Even Reid liked Soapy personally, but he had no doubt that he was a menace to real, as opposed to Smith's, law and order. On July 4, Independence Day, 1898, it seemed

Soapy Smith, bearded and benign, with some of his choicest Skagway con-men

that nothing could shake Soapy's throne, but on the 8th he was dead. The cause was an unlikely one. A young miner, J. D. Stewart, was on his way home with a pouch containing $2,800 in gold, the result of two years' work. Despite warnings by his friends, he fell into the gang's clutches, lost it all, and wandered, like so many before, through the streets lamenting his ill-fortune. A lawman in Soapy's pay could only suggest a return trip to the Klondike.

A human flood was expected out of Dawson any day now with the Yukon open after the long winter, and those merchants who heard the lamentations of Stewart were suitably shocked. Word was sent to Frank Reid, who, with his friend Captain Sperry and a Major Tanner, rapidly reorganised their vigilantes.

Soon the whole town was humming with excitement, but Soapy, whose bravery was never in question, marched out in front of an angry crowd, a revolver showing in outline in each pocket, and shouted: 'You're a lot of cowardly rope-pulling sons of bitches. Now come on! I can lick the whole crowd of you.' They dispersed.

But down from Dyea came Judge Sehlbrede who confronted Smith with an ultimatum that Stewart's poke must be returned by 4 p.m. Soapy alleged that Stewart had lost it gambling and that perhaps he might be able to restore it.

The deadline came and went, and it was ominous for Soapy that another crowd did not melt away. It was Soapy who turned on his heels. Tension went on mounting, as some of Soapy's weaker brethren slipped out of town, and the climax came that night on the end of Jumeau Wharf.

Billy Saportas, Soapy's tame reporter, had sent him a note that the crowd at a vigilante meeting was angry and that he had better act quickly, so Soapy pocketed a Colt .45, slung a Winchester round his shoulder, and went to 'teach these damned sons of bitches a lesson.' Only a dozen of his men were there behind him. The rest had fled.

Soapy knew whom he had to kill–Frank Reid, who stood ahead of the rest, a revolver in his pocket. He ordered Soapy to halt.

'Damn you, Reid,' said Soapy, 'you're at the bottom of all my troubles. I should have got rid of you three months ago.' Then he levelled his rifle at Reid's head. Reid seized its muzzle with his left hand and pulled it downwards, reaching for his revolver with his right.

'Don't shoot!' cried Soapy, unnerved at last, but Reid squeezed the trigger. Tragically the cartridge was faulty and Reid received the bullet in his groin. He fired again at Smith, who dropped dead, a bullet in his heart. As Reid fell, he fired a third time, hitting Soapy above the left knee.

Now six or seven of Soapy's men, their guns drawn, looked like avenging their dead leader, but the sight of the determined vigilantes and the clear hostility of everyone

Superintendent Sam Steele and his handful of Mounties kept the Klondike virtually free from crime. He was nicknamed 'The Lion of the Yukon'

else made them put away their guns and run, while Reid called out: 'I'm badly hurt, boys, but I got him first.'

He was given three hearty cheers then was carried away to the town hospital, still elated at his feat. Cruelly, he was to linger on in agony for twelve days.

Meanwhile, Soapy's gang found themselves trapped in Skagway, except for Bowers, Tripp and Foster, who had managed to escape towards the White Pass, to be rounded up later by vigilantes. In all, 26 were caught and placed in a jail outside which a crowd howled for their blood, their feelings being roused even more by the knowledge that their champion was dying. Amazingly, all the prisoners left town alive, proving that law and order had come to stay. Eleven were sent to Sitka where they were sentenced to between one and ten years, and the rest were put on a Seattle-bound boat.

Neither the Baptist nor the Methodist ministers of Skagway wished to bury Soapy, but the Presbyterian, Rev. Sinclair agreed to, with a vigilante to protect him. The king of the con men was found to have remarkably little money left, having squandered his wealth, much of it on trying to lick the one game he never mastered, faro. And he had given much money away.

Frank Reid was operated on, but without success. As he watched the operating table being prepared, he puffed at a cigar and said: 'If that bullet hasn't punctured my insides and if blood-poisoning doesn't set in and if you damned doctors don't butcher me, I'll have a chance for life yet.'

They gave him the biggest funeral in Skagway's history and buried him near Soapy Smith's grave. On the marble slab that marks it is the inscription: 'He gave his life for the honor of Skagway.'

Just a month before the downfall of Soapy Smith, the main mass of stampeders reached Dawson, having waited along Lakes Lindermann and Bennett for the ice on the Yukon to break. Most of them had come across the Chilkoot, which for all its terrors and the fact that no animal except man could carry baggage across it, was the best way to reach the Klondike.

It began at Dyea, near Skagway and almost its twin in appearance, and was a town full of Indians ready to carry baggage over the pass at a competitive price, except on Sundays: they were strict Presbyterians and had a business sense worthy of a Scots fellow-religionist. From Dyea to Lake Lindermann was less than 30 miles, and the section before the foot of the Chilkoot began smoothly enough through a pleasant country criss-crossed by a river, until gradually the sides of the route were seen to be littered with discarded baggage of every kind. The going got worse until the base of the pass was reached at Sheep Camp, once the headquarters of some mountain sheep hunters. Now the stampeder could see the grim four-mile climb ahead of him, and the endless line of men heading for the summit.

Sheep Camp was a town of sorts, consisting of a shack for professional packers and a few other tents, sheds and 'hotels'. Women were available for $5, a plate of bacon, beans with tea to wash it down for half as much. It was here that horses were abandoned by their owners who found that they could not use them–and abuse them–any further. Most were shot and lay under the snow that fell, often continuously, from the autumn of '97 till the spring of '98.

On the four-mile climb, which reached nearly 35 degrees, the stampeders did what came to be known as 'The Chilkoot Lock-Step' as they moved ever upward. There was nothing to frighten an experienced climber on the Chilkoot, but very few of the 22,000 were mountaineers, the majority of them were undernourished or sick or both, and all were utterly exhausted. In their overheavy furs and woollen garments they either roasted or froze, while the terrible mountain winds often turned to blizzards which halted all progress. They stank because they lived in their clothes.

Most of them were too poor to hire carriers, which meant that getting a ton of goods across the pass was a matter of journey after journey to the top, for the average climber could only manage a 50-pound pack at a time. It could take three months of shuttling back and forth, perhaps 40 climbs.

The last 150 feet were the worst, and steps had been hacked out with axes by enterprising climbers who then charged others for the privilege of using them. There were even some resting shelves along with a growing number of steps, but the climber who dared take a rest was liable to have to wait a full day before getting his place back in the Lock-Step.

Naturally, the packers were under no great strain, being used to the climb and in perfect condition, as was the young American who took a 125-pound plough up the final slope. One Indian reached the top with a 350-pound barrel on his back, while at one time or another up the Chilkoot went pianos, sections of steamboats and whisky hopefully concealed well enough to be got through the Canadian Customs post.

The ton of supplies required of each stampeder included the food that was to prevent starvation in the Klondike in 1898, plus every sort of tool and equipment, from pans to tents, and clothing. Dumps were marked with poles, after which stampeders descended at speed, usually on their backsides. However, by the Spring of '98, a clever aerial tramway had been installed, consisting of some 14 miles of copper-steel cable that was supported by huge tripods with steam engines at both ends. It could move goods in 300-pound carloads at a rate of a carload a minute.

By then, most of the men of '98 had climbed the Chilkoot

the hard way, straining hour after hour, too short of breath to curse, but sometimes collapsing, occasionally weeping with frustration, always regaining the will to fight on upwards. On this grimmest of journeys some died when avalanches struck the mountain and buried them alive. The rest fought on, not so much because of their gold fever, but because they had set themselves a task and would not be beaten.

A city of freight grew at the top of the pass, much of which was to remain buried until the thaw came. And at the top were the Mounties in their buffalo coats, a sentry on duty at a Maxim gun, and a Union Jack flying. They endured everything that Nature could hurl at them that winter in their small hut and outside it, checking the 22,000 through, Americans, Britons from all over the Empire, representatives of nearly every nation, women included, dancing girls destined for Dawson, wives, even an old German woman of 70. And an English nobleman went up the Chilkoot, complete with valet, while many went not to find gold but sensibly to start a business, including a newsboy with papers to sell to news-hungry miners.

Soapy Smith's men were there, selling seats for the travel-weary, selling places by a fire, and, amazingly, setting up various con games on the trail, which drew players eager for a break; but the gang never got past the Mounties.

The Chilkoot trail ended at Lake Lindermann, where thousands began to build boats beside its shore. The majority, however, pressed on over the ice to reach the larger Lake Bennett, where those who had come over the White were building their boats. A dangerous, boulder-filled canyon linked the two lakes, and stories of what had happened to early travellers in '97 frightened most stampeders out of attempting a trip through the rapids of the canyon when the ice finally broke. Some were even further forward at Lake Tagish, which would get them down to Dawson that much sooner.

So it came about that some 30,000 stampeders were busily constructing more than 7,000 boats over the 60 miles from Lakes Lindermann to Tagish, a statistic which must be rapidly broken down into the human drama that it was. Lake Bennett's tent city was the biggest in the world and surrounded the entire lake. The photograph of just a portion of it shows that the word 'city' is justified. The citizens' troubles were by no means over, but at least they had learnt something of the art of survival in the mountains, though with the protective beards, their skin smeared with charcoal to stop sunburn and their simple masks to protect their eyes from fierce snow glare, they presented an appearance made even more stark by loss of weight and worn clothes.

The long months beside the lakes went smoothly because of the presence of Steele and his men.

One of the originals who had crossed the Plains in '73, and with a long and distinguished career behind him, 'Steele of the Mounted', 'The Lion of the Yukon', both titles admiringly bestowed, reached heights in '98 rare even for a Mountie. He ruled the Klondike along military lines, but with humanity and understanding, and it was sober truth that a sack of nuggets left on the trail for two weeks would be there when its owner returned.

It was he who saw to it that every stampeder had a year's food supply, he who ordered that each boat should have a serial number on its bow and that his men took particulars of each occupant, sending details along the river, so that if a boat failed to appear, a search could be organised. He and his men, more of whom were Britons than Canadians, including 'younger sons of well-to-do families, seeking adventure and service in the outposts of empire,' as Pierre Berton has characterised them in his generous tribute to men who 'seldom raised their voices, almost never drew a gun, and rarely had to give an order twice.'

These were the men who kept Soapy Smith's gang out. One did once seem to have got through, but not for long. 'I'm an American citizen,' he told Steele after he had been caught with some of the tools of the con man's trade. 'I'll have you know you can't lock up a United States citizen and get away with it. My God, sir! The Secretary of State himself shall hear about it!'

'Well,' said the Lion of the Yukon, 'seeing you're an American citizen, I'll be very lenient, I'll confiscate everything you have and give you half an hour to leave.' Shortly afterwards, the intruder was being marched back up the trail a full 22 miles to the summit with a Mountie at his heels barking at him to 'step lively.'

On May 29, 1898, the ice cracked and two days later 7,124 craft set out, some being merely three logs bound together and others being actual coffins. There were 500 miles to go but at last they were on their way, the incredible armada with its humans, its animals–oxen, horses, dogs–its precious belongings. And above them, looking down anxiously on the scene, was the olympian figure of Sam Steele, who knew just how little those below knew about sailing. The first night out the wind dropped and the thousands of stampeders rested on a lake far from all their homes, yet with the worst behind them.

There was, however, one terrible gauntlet still to be run, for 50 miles on was a ferocious stretch of water, Miles Canyon, 100 feet wide and 50 feet deep, through which the Lewes River roared. In the canyon was a whirlpool followed by rapids. In the first days of the river rush, 150 craft were destroyed in the canyon and ten men were drowned, so thousands of other boats held back, their crews frightened and demoralised. Others huddled at the far end of the canyon, all their belongings lost. The bottleneck was total, the chaos appalling, but at that moment Sam Steele arrived like a latter-day Moses, to lead his people out of adversity.

He calmed the mob, and ordered that no boat must shoot the rapids unless first checked by his Corporal Dixon 'who thoroughly understands this work', on pain of a heavy fine. All women and children were told to walk the five miles to below the rapids, and all boats were to be steered by competent men, approved by the corporal. From that moment onwards, scarcely another boat was lost, while a bright young man named Macaulay soon had a tramway in action, complete with horse-drawn cars, charging $25 dollars to take a boat below the rapids.

The boats raced on down the river after the nightmare of the canyon had been ended. Some men had already reached Dawson across the ice, including one smart operator who arrived with 2,400 eggs which he promptly sold for $3,600. Then on June 8, the armada was sighted and within a matter of days Dawson was a booming gold rush town. By July 1, it even had two banks. Boats which had at last managed to sail up the Yukon had reached the town and small steamers had sailed down to Dawson, after being taken over the Chilkoot piece by piece and reconstructed.

Dawson had had a worrying Spring, because it had seemed that the Yukon might flood the low-lying land on which it had been built. The river started to subside three days before the armada arrived, when the business section was under five feet of water. When the stampeders reached town the streets were an ocean of mud. Tents that had housed thousands in camps along the trails were raised yet again while the stampeders set to work to build more permanent structures, when they were not trying to find friends and partners in the chaos of a booming town desperately short of street addresses.

There were around 5,000 people in Dawson in the summer of '97, but no one knows just how many were there a year later. It may have varied between 25,000 and 40,000. It was clear soon after arrival that the men of '98 were too late, that the best claims had already been staked. There were those who were crushed by the news, others who followed every wild rumour of a strike, still more who wandered around listlessly, the magnificent virility and willpower that had made them endure such hardships now turned to a meaningless inertia. They became sightseers.

Yet many seem to have been happy enough, for they had reached Eldorado and simply to have got to it was a matter for self-congratulation. Many kept their Chilkoot clothes on like uniforms all that hot summer, even remaining shaggily bearded. Many spent their days selling off, or trying to sell off, enough of their statutory ton of goods to pay for their journey home, so the town became a giant street market by day, while by night Dawson was like so many of the mining towns we have visited in the course of this book, but with one unique quality. In the whole of that legendary year of '98, there was not a single murder in the town. Men could even leave their cabin doors unlocked with their gold inside.

Steele took over from Constantine in June, the latter being given a notable tribute by the veteran miners who knew his worth. The man who had first given them law and order and made not a penny himself, was presented with a silver plate on which were $2,000 of nuggets. Steele, coming in on the flood tide of the stampeders, drew the line at obscenity, cheating and disorderly conduct, but had no objection to the sudden upsurge of gambling halls, saloons and brothels as long as public order was not threatened. His only restrictions on the ocean of liquor that flooded into Dawson were to ban its sale to minors and to forbid the employment of children in saloons. Like the best type of army officer that he so resembled he ruled firmly but fairly, with a true understanding of his charges.

Firearms, of course, were forbidden in Dawson, with offenders being either expelled, given a 'blue ticket' to leave town, or made to saw wood to heat government buildings, a very unpopular punishment, especially as it had to be sawn to fit the stoves. One gambler sneered when Steele fined him $50. 'Is that all?' he asked. 'I've got that in my vest pocket.' '... and sixty days on the woodpile,' continued Steele. 'Have you got that in your vest pocket?'

The Sabbath was a day of calm, and some theatre owners at least managed to present 'sacred concerts' of tableaux from the scriptures, though one evening a buxom dance hall queen named Caprice topped the bill stark naked except for tights, slippers and a huge cross which she clasped to herself most suggestively.

1898 saw one new technique used in the Klondike by the minority actually engaged in mining. It will be remembered that thawing the ground by 'burning' was a slow process, but C. J. Berry, a hard-working ex-fruit farmer from California, who kept the $1½ million he made in the Klondike and lived until 1930 worth far more, found a better method of thawing. He had noted that steam escaping from an engine exhaust had thawed a hole, which led him to experiment with a rubber hose. The result was his 'steam point'. Via a boiler on the surface, steam was sent through a hose to an iron pipe, five or six feet in length. The pipe was pointed at its lower end and had an orifice for the steam to be applied to the ground. Once inserted into the frozen gravel, it was pushed steadily forward by gentle hammer taps. Burning had not even been totally effective in summer because of wet ground, and Berry's process, widely used and soon improved, was anyway far faster and more efficient.

By the summer of 1899 companies were buying up claims in the Klondike and mechanisation was taking over from individuals. The 'poor man's rush' was finished. Just how much gold was extracted is difficult to estimate as the Canadian Government's 10% royalty on the gross output, even set against a $2,500 annual allowance for the cost of extraction, undoubtedly led to some concealment of gains. In 1900, when most miners had gone home or headed for a new strike at Nome on the western coast of Alaska, gold production was at its height in the Klondike, 1,350,000 fine ounces were extracted, worth $22,250,000, and between 1897 and 1905 the total was worth $100 million. Since 1897, the Klondike has produced around $300 million and work goes on to this day, but Dawson, for a little over a year the last fabulous gold mining boom town, now houses some 800 people. Though trains go through the White Pass, the Chilkoot remains majestically silent. The camps have gone, Dyea has gone. As for Alaska, now a state of the Union, it remains a gold-producing area, though black gold is now its main concern in the oil-conscious 1970s. Nome had some brief years of gaudy glory in the old sense, but the gold rushes may be said to have ended in the unforgettable summer of '98.

In harsh, practical terms, the Klondike, as will have become obvious, was not one of the greatest gold rushes, but this has been a book primarily about people, and in terms of the human spirit it was the greatest of them all. Not that the practical results were negligible. The Klondike was mainly responsible for the opening up of Alaska because of the strikes that followed it, while a number of cities benefited hugely, most notably Seattle, Edmonton and Vancouver. Western Canada's opening up increased space and there was no more talk of a slump in America's Northwest. More impressive still is the sheer number of Klondikers who sprang into prominence in America and, especially, in Canada, though their names are less known now. Remarkable too is the fact that a machine-gun battalion of Klondike pioneers became the most decorated Canadian unit in the First World War, more than 60% being awarded medals for bravery. In today's cynical climate such a statistic may carry less weight, but it is significant because it affords striking proof of the determination that carried men over the Chilkoot, that made obstacles in later life, however severe, seem there to be conquered. Long after they had made their epic journey, they still, with Tennyson's Ulysses, were 'strong in will to strive, to seek, to find, and not to yield.'

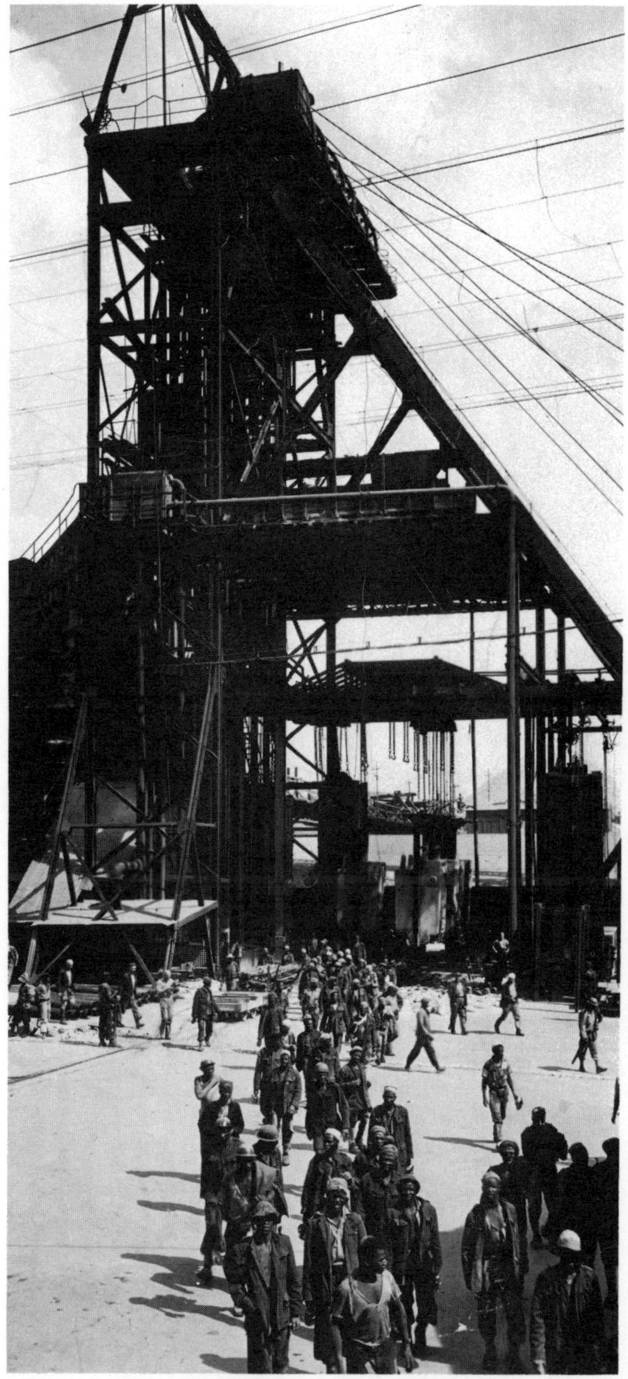

A gold mine on South Africa's Rand. Almost from the beginning, machinery dominated the fields in South Africa

Index

Acknowledgements

The author's grateful thanks are due to Merrilee Dowty, Elaine Gilleran and the other ladies of the Wells Fargo Bank History Room in San Francisco for their help in supplying so many fine photographs, also two other Californian friends, Andrew Rolle and Richard Dillon, for advice and encouragement. Katherine Halverson of Wyoming was, as always, most helpful, and Australian, New Zealand and South African librarians in London and the staff of the Public Archives of Canada and the Provincial Archives of Victoria, B.C. dealt with all the author's requests courteously and speedily.

Books used as sources
As well as the books and diaries mentioned in the text, the author found the following particularly useful and in most cases entertaining: *The Bonanza West* (Norman, 1963) by William Greever; *Klondike* (London, 1960), Pierre Berton's classic on the subject; *Gold Fever* (Sydney, 1967) edited by Nancy Keesing; *Gold! Gold! Gold!* (London, 1970) by Eric Rosenthal, who describes South Africa's rushes; and W. H. Morrell's scholarly and sound *The Gold Rushes* (London, 1968). Time-Life's visually stunning *The Forty-Niners* (1974) also has a fine text, while *The Golden Frontier* (Austin, 1962) is Herman Reinhart's own account of his adventures in the not-very-well-known rushes to Oregon and British Columbia. For those who wish to relate the American and Australian rushes more fully to their historical settings, good general works include John Hawgood's *The American West* (London, 1967); *Australia: History and Horizons* (London, 1971) by Roderick Cameron, and *A Concise History of Australia* (London, 1968) by Clive Turnbull.

Photo credits
The author and publishers would like to thank the following sources for providing the photographic illustrations reproduced in this book. Credits are given spread by spread.

Pages	
1	Provincial Archives, Victoria, B.C.
2–3	The Commissioner, Royal Canadian Mounted Police
4–5	Wells Fargo Bank History Room
6–7	John Frost Newspapers
8–9	Society of California Pioneers; The Bancroft Library
10–11	Robin May; Mansell Collection
12–13	Mansell Collection
14–15	Robin May; Mansell Collection
16–17	Wells Fargo Bank History Room
18–19	Mansell Collection; U.S. National Archives
20–21	Handbook of Early Advertising Art/Clarence P. Hornung;
22–23	Handbook of Early Advertising Art/Clarence P. Hornung
24–25	Handbooks of Early Advertising Art/Clarence P. Hornung; U.S. National Archives
26–27	Kansas State Historical Society; Wells Fargo Bank History Room; Handbook of Early Advertising Art/ Clarence P. Hornung
28–29	Anne-Marie Erhlich
30–31	Handbook of Early Advertising Art/Clarence P. Hornung: Arizona Historical Society
32–33	Robin May
34–35	Wells Fargo Bank History Room; Robin May
36–37	Robin May
38–39	John Frost Newspapers; Handbook of Early Advertising Art/Clarence P. Hornung; U.S. National Archives
40–41	U.S. National Archives
42–43	Anne-Marie Erhlich
46–47	Australian News; Art Gallery of New South South Wales; Mansell Collection
48–49	Anne Marie–Erhlich; Anne–Marie Erhlich/ Hotterman
50–51	Anne Marie-Erhlich; Mansell Collection
52–53	Anne Marie-Erhlich
54–55	Rank Film Distributors courtesy of the Stills Library, National Film Archive, London
58–59	Australian Information Service
60–61	Mansell Collection
62–63	Mansell Collection
64–65	Mansell Collection
66–67	Wells Fargo Bank History Room
68–69	Wells Fargo Bank History Room
70–71	Robin May
72–73	Wells Fargo Bank History Room
74–75	R.K.O. Film Corporation courtesy of the Stills Library, National Film Archive, London
76–77	U.S. National Archives
78–79	U.S. National Archives
80–81	Minnesota History Society
82–83	Kansas State Historical Society; Provincial Archives, Victoria, B.C.; Robin May
84–85	U.S. National Archives
86–87	U.S. National Archives; Smithsonian Institute National Anthropological Archives
88–89	U.S. National Archives
90–91	Montana Historical Society; Wells-Fagliano
94–95	Mansell Collection; Wells Fargo Bank History Room (insert); Mary Evans Picture Library (and insert)
96–97	Mary Evans Picture Library
100–101	John Frost Newspapers
102–103	John Frost Newspapers; Public Archives of Canda
104–105	The Commissioner, Royal Canadian Mounted Police
106–107	The Public Archives of Canada
108–109	Louis Simonin's *La Vie Souterrane* Robin May
110–111	Western Collection, Denver Public Library
112–113	The Commissioner, Royal Canadian Mounted Police
116–117	Mansell Collection

Front cover: Wells Fargo Bank History Room
Back Cover: The Public Archives of Canada